High Precision X-Ray Measurements

High Precision X-Ray Measurements

Special Issue Editor

Alessandro Scordo

MDPI • Basel • Beijing • Wuhan • Barcelona • Belgrade

Special Issue Editor
Alessandro Scordo
INFN Laboratori Nazionali di Frascati
Italy

Editorial Office
MDPI
St. Alban-Anlage 66
4052 Basel, Switzerland

This is a reprint of articles from the Special Issue published online in the open access journal *Condensed Matter* (ISSN 2410-3896) in 2019 (available at: https://www.mdpi.com/journal/condensedmatter/special_issues/X_Ray)

For citation purposes, cite each article independently as indicated on the article page online and as indicated below:

LastName, A.A.; LastName, B.B.; LastName, C.C. Article Title. *Journal Name* **Year**, *Article Number*, Page Range.

ISBN 978-3-03921-317-7 (Pbk)
ISBN 978-3-03921-318-4 (PDF)

Contents

About the Special Issue Editor

Alessandro Scordo studied physics at the Sapienza University of Rome from 2001 to 2006, where he graduated with a degree in condensed matter physics, having completed an experimental thesis on semi-insulator materials. He received a PhD in 2011 from the INFN Laboratories of Frascati, where, after six more years of postdoc research, he is presently working on the SIDDHARTA-2 experiment. His research interests include nuclear physics, particle physics, solid-state and radiation detectors, and data acquisition systems, and he has been part of the SIDDHARTA, SIDDHARTA-2, and AMADEUS collaborations at LNF, the VIP collaboration at LNGS, the HADES collaboration at GSI, and the E15, E17, E31, E45, E62, and E72 collaborations at JPARC/RIKEN. In 2015, he was named as one of the six "Young Research Grant" recipients of the INFN CSN5, working on a mosaic crystal-based von Hamos spectrometer for extended sources. His results have been published in internationally renowned peer-reviewed journals (169 publications to date, with 1029 citations, and h = 14).

Editorial

High Precision X-Ray Measurements

Alessandro Scordo

Istituto Nazionale di Fisica Nucleare–Laboratori Nazionali di Frascati (LNF-INFN), Frascati, 00044 Roma, Italy; alessandro.scordo@lnf.infn.it

Received: 17 June 2019; Accepted: 22 June 2019; Published: 25 June 2019

Keywords: X-ray; XAS; XRF; multidisciplinarity; X-ray source facilities; material investigation; graphite crystals

Since their discovery in 1895, the detection of X-rays has had a strong impact and various applications in several fields of science and human life. Kiloelectronvolt (keV) photons are indeed the leading actors in a large variety of physics fields, and impressive efforts have been and are being done to develop new type of detectors and new techniques, aiming to obtain higher precision measurements of their energy, position and polarization.

Historically, new detectors and monochromators for synchrotron radiation high precision measurements were first made in Europe at the Frascati laboratories by a Italian–French team in the sixties, using the 1 GeV electron synchrotron [1]; a decade later, these efforts were followed by the Italian team of Balzarotti and Bianconi [2].

Applications of X-ray measurements are relevant in fundamental and nuclear physics, biophysics, medical research, molecular and surface structure of materials studies.

The aim of this special issue is to have a global overview, from these communities and research fields, of the most recent developments in X-ray detection techniques and their impact.

To accomplish this aim, the published papers provide high quality research results, giving an insight into the hot topics and the main applications of X-ray photons.

In the paper of Andreeva et al. [3], a detailed description on the possible use of Mössbauer spectroscopy in material surface investigations is given, with particular emphasis on the importance of the polarization analysis in reflectivity experiments; the authors show the advantages of the polarization selection to deliver high quality data, providing a better interpretation of the magnetic ordering in multilayer films.

The properties of materials, and in particular their crystalline phase, is investigated in the paper of Macis et al. [4], where the structural changes of MoO_3 thin films upon annealing at different temperatures are investigated using X-ray Absorption Spectroscopy (XAS). The presented results are very promising and suggest the possibility of using such material as a hard, protective, transparent and conductive material in different technologies. It represents the first important advancement for many applications, in particular for the development of compact radio frequency (RF) accelerating devices made of copper.

Concerning biological samples analysis, the first high precision Ca K-edge X-ray Absorption Near Edge Spectroscopy (XANES) measurement of CaATP molecules was performed using the Slac storage ring of Stanford University [5], followed by the high precision X-ray spectroscopy measurements on MnATP molecules at the first 1.5 GeV European synchrotron radiation facility using a large storage ring (ADONE, INFN, Laboratories of Frascati) [6].

The usage of XAS technique on biological samples is also the main subject of the paper from E. De Santis et al. [7], with a particular effort toward the possibility of performing XAS measurements on very diluted samples, with an absorber concentration at the micromolar level. The reported measurements, focused on the Cu(II)-ProIAPP (islet amyloid polypeptide) complexes under near-physiological and equimolar concentrations of Cu(II) and peptide, may have a strong medical impact related to the cell death in type 2 diabetes mellitus, with important consequences on human health.

Another possible medical application of keV photons is described in the paper of Makek et al. [8], where the performances of a single-layer scintillator pixel detector are reported; the obtained results demonstrate the possibility of detecting Compton gammas with an energy and timing resolution comparable to those achieved in photo-electric absorption measurements. In addition, the authors show how the detection and full reconstruction of such gammas is very interesting and promising in medical imaging, with particular emphasis on the proton emission tomography (PET) technique, where measurement of polarization correlations of annihilation photons may improve the required sensitivity leading to a reduction of the needed number of electronic channels and of the overall costs.

X-ray measurements are a perfect tool for testing fundamental principle of physics, like the Pauli exclusion principle (PEP); this is reported and described in the paper by K. Piscicchia et al. [9], where the results of the measurements performed by the Violation of the Pauli exclusion principle (VIP) experiment at the INFN underground Laboratories of Gran Sasso are presented. The photons coming from the atomic transitions of a copper sample are detected and analysed, looking for the possible existence of PEP violating electrons producing an anomalous X-ray signal corresponding to the transition from the 2p level to the 1s level when this last one is already occupied by two electrons. The authors provide the most recent limit on the PEP violation probability, together with a detailed explanation of the performed analysis.

X-rays coming from atomic transitions can be also used, in nuclear physics, to investigate the low energy interaction involving strange quarks by measurements of kaonic atoms; this possibility is explored and reported in the paper from Curceanu et al. [10], which provides an overview of the measurements of kaonic atoms performed at the Double Anular Φ-factory for Nice Experiments (DAΦNE) collider at LNF-INFN and on the X-ray detectors used in the experiments, like charged couple devices (CCDs) and silicon drift detectors (SDDs).

These latter devices are, nowadays, the best performing large area spectroscopic detectors in terms of energy resolutions, and their faster timing capability with respect to CCDs make them suitable for measurements in high background environments. The characterization of a set of SDDs, produced by Fondazione Bruno Kessler (FBK, Trento, Italy) and used in the SIDDHARTA-2 (Silicon Drift Detector for Hadronic Atom Research by Timing Application) experiment at the DAΦNE collider, in terms of stability and linearity is given by Miliucci et al. [11]; in this work, a linear response within 1 eV is reported for photons in the energy range 4–12 keV, for typical energy resolutions of 120 eV @ 6 keV Full Width at Half Maximum (FWHM).

A factor 50 improvement in the energy resolution, with respect to SDDs, can be achieved with the Bragg spectroscopy technique, with the drawback of waiving the large acceptances and efficiencies typical of solid state devices. The possibility of pushing this technique toward millimetric and isotropic sources, instead of the standard micrometric collimated ones, is explored by the VOXES (high resolution VOn hamos X-ray spectrometer using HAPG for Extended Sources) project and reported in the paper by Scordo et al. [12], where a new Bragg spectrometer with graphite mosaic crystals in the Von Hamos configuration is proposed. In this paper, results of the measurements of the copper and iron K$\alpha_{1,2}$ performed with different mosaicity crystals are discussed, investigating the effect of both the mosaicity and the crystal thickness on the energy resolution when keeping the source dimension at the millimetre level. The obtained results are very promising and trigger the possibility of using such spectrometers for nuclear physics experiments, as well as material investigations and trace metals identification, with several important applications in other fields.

The above-mentioned research is based on the usage, as diffraction crystals, of mosaic highly annealed pyrolitic graphite (HAPG) and highly oriented pyrolitic graphite (HOPG) crystals. These objects are experiencing a very rapid expansion on the market, due to their extremely interesting mechanical properties and physical properties, like a small lattice constant, low thermal expansion coefficient and the possibility of depositing them on a substrate of any geometrical shape. These characteristics, together with the HAPG and HOPG production mechanisms, are presented by one of the main graphite crystal producers, the OPTIGRAPH GmbH company, in the paper by Grigorieva et al. [13].

All the above-mentioned papers clearly suggest how the possibility of having access to X-ray sources is mandatory to perform high quality research in various fields.

One possibility is given by the DAΦNE-Light beam facility at INFN Laboratories of Frascati, described in detail by A. Balerna in [14]; in particular, the soft X-ray DXR1 beamline is presented, together with a description of the typical XAS applications that can be performed on site.

The DXR1 beamline started delivering beamtime to users at the end of 2004; more recently, a proposal for building a free electron laser (EuPRAXIA@SPARC_LAB) at the INFN Laboratories of Frascati is under consideration. The technical details and the beam characteristics achievable in the main experimental lines are the main subject of the paper by A. Balerna et al. [15].

All the papers included in this Special Issue contribute to underlining the importance of X-ray detection for a very broad range of physics topics; most of these topics are covered by the published works and several others are mentioned in the paper references, providing an interesting and very useful overview from these different communities and research fields, of the most recent developments in X-ray detection and their impact on fundamental research and societal applications.

The readers of this special issue may also find other review papers published in other special issues of Condensed Matter, focused in particular on synchrotron radiation techniques [16,17] very interesting.

Acknowledgments: I express my thanks to the contributing authors of this Special Issue, and to the journal Condensed Matter and MDPI for their support during this work.

Conflicts of Interest: The author declares no conflict of interest.

References

1. Jaeglé, P.; Missoni, G.; Dhez, P. Study of the Absorption of Ultrasoft X Rays by Bismuth and Lead Using the Orbit Radiation of the Frascati Synchrotron. *Phys. Rev. Lett.* **1967**, *18*, 887. [CrossRef]
2. Balzarotti, A.; Bianconi, A.; Burattini, E.; Grandolfo, M.; Habel, R.; Piacentini, M. Core transitions from the Al 2p level in amorphous and crystalline Al_2O_3. *Phys. Status Solidi B* **1974**, *63*, 77–87. [CrossRef]
3. Andreeva, M.; Baulin, R.; Chumakov, A.; Kiseleva, T.; Rüffer, R. Polarization Analysis in Mössbauer Reflectometry with Synchrotron Mössbauer Source. *Condens. Matter* **2019**, *4*, 8. [CrossRef]
4. Macis, S.; Rezvani, J.; Davoli, I.; Cibin, G.; Spataro, B.; Scifo, J.; Faillace, L.; Marcelli, A. Structural Evolution of MoO_3 Thin Films Deposited on Copper Substrates upon Annealing: An X-ray Absorption Spectroscopy Study. *Condens. Matter* **2019**, *4*, 41. [CrossRef]
5. Bianconi, A.; Doniach, S.; Lublin, D. X-ray Ca K edge of calcium adenosine triphosphate system and of simple Ca compunds. *Chem. Phys. Lett.* **1978**, *59*, 121–124. [CrossRef]
6. Belli, M.; Scafati, A.; Bianconi, A.; Mobilio, S.; Palladino, L.; Reale, A.; Burattini, E. X-ray absorption near edge structures (XANES) in simple and complex Mn compounds. *Solid State Commun.* **1980**, *35*, 355–361. [CrossRef]
7. De Santis, E.; Shardlow, E.; Stellato, F.; Proux, O.; Rossi, G.; Exley, C.; Morante, S. X-Ray Absorption Spectroscopy Measurements of Cu-ProIAPP Complexes at Physiological Concentrations. *Condens. Matter* **2019**, *4*, 13. [CrossRef]
8. Makek, M.; Bosnar, D.; Pavelić, L. Scintillator Pixel Detectors for Measurement of Compton Scattering. *Condens. Matter* **2019**, *4*, 24. [CrossRef]
9. Piscicchia, K.; Amirkhani, A.; Bartalucci, S.; Bertolucci, S.; Bazzi, M.; Bragadireanu, M.; Cargnelli, M.; Clozza, A.; Curceanu, C.; Del Grande, R.; et al. High Precision Test of the Pauli Exclusion Principle for Electrons. *Condens. Matter* **2019**, *4*, 45. [CrossRef]
10. Curceanu, C.; Amirkhani, A.; Baniahmad, A.; Bazzi, M.; Bellotti, G.; Berucci, C.; Bosnar, D.; Bragadireanu, M.; Cargnelli, M.; Clozza, A.; et al. X-ray Detectors for Kaonic Atoms Research at DAΦNE. *Condens. Matter* **2019**, *4*, 42. [CrossRef]

11. Miliucci, M.; Iliescu, M.; Amirkhani, A.; Bazzi, M.; Curceanu, C.; Fiorini, C.; Scordo, A.; Sirghi, F.; Zmeskal, J. Energy Response of Silicon Drift Detectors for Kaonic Atom Precision Measurements. *Condens. Matter* **2019**, *4*, 31. [CrossRef]

12. Scordo, A.; Curceanu, C.; Miliucci, M.; Sirghi, F.; Zmeskal, J. Pyrolitic Graphite Mosaic Crystal Thickness and Mosaicity Optimization for an Extended Source Von Hamos X-ray Spectrometer. *Condens. Matter* **2019**, *4*, 38. [CrossRef]

13. Grigorieva, I.; Antonov, A.; Gudi, G. Graphite Optics—Current Opportunities, Properties and Limits. *Condens. Matter* **2019**, *4*, 18. [CrossRef]

14. Balerna, A. DAFNE-Light DXR1 Soft X-Ray Synchrotron Radiation Beamline: Characteristics and XAFS Applications. *Condens. Matter* **2019**, *4*, 7. [CrossRef]

15. Balerna, A.; Bartocci, S.; Batignani, G.; Cianchi, A.; Chiadroni, E.; Coreno, M.; Cricenti, A.; Dabagov, S.; Di Cicco, A.; Faiferri, M.; et al. The Potential of EuPRAXIA@SPARC_LAB for Radiation Based Techniques. *Condens. Matter* **2019**, *4*, 30. [CrossRef]

16. Campi, G.; Bianconi, A. Evolution of Complexity in Out-of-Equilibrium Systems by Time-Resolved or Space-Resolved Synchrotron Radiation Techniques. *Condens. Matter* **2018**, *4*, 32. [CrossRef]

17. Baccolo, G.; Cibin, G.; Delmonte, B.; Hampai, D.; Marcelli, A.; Di Stefano, E.; Macis, S.; Maggi, V. The Contribution of Synchrotron Light for the Characterization of Atmospheric Mineral Dust in Deep Ice Cores: Preliminary Results from the Talos Dome Ice Core (East Antarctica). *Condens. Matter* **2018**, *3*, 25. [CrossRef]

Article

Polarization Analysis in Mössbauer Reflectometry with Synchrotron Mössbauer Source

Marina Andreeva [1,*]**, Roman Baulin** [1]**, Aleksandr Chumakov** [2,3]**, Tatiyana Kiseleva** [1] **and Rudolf Rüffer** [2]

[1]   Faculty of Physics, M.V. Lomonosov Moscow State University, 119991 Moscow, Russia;
      baulin.roman@physics.msu.ru (R.B.); kiseleva.tyu@gmail.com (T.K.)
[2]   ESRF-The European Synchrotron, CS 40220, 38043 Grenoble CEDEX 9, France; chumakov@esrf.fr (A.C.);
      rueffer@esrf.fr (R.R.)
[3]   National Research Centre "Kurchatov Institute", Pl. Kurchatova 1, 123182 Moscow, Russia
*   Correspondence: Mandreeva1@yandex.ru; Tel.: +07-903-712-0837

Received: 1 October 2018; Accepted: 2 January 2019; Published: 8 January 2019

Abstract: Polarization selection of the reflected radiation has been employed in Mössbauer reflectivity measurements with a synchrotron Mössbauer source (SMS). The polarization of resonantly scattered radiation differs from the polarization of an incident wave so the Mössbauer reflectivity contains a scattering component with 90° rotated polarization relative to the π-polarization of the SMS for some hyperfine transitions. We have shown that the selection of this rotated π→σ component from total reflectivity gives an unusual angular dependence of reflectivity characterized by a peak near the critical angle of the total external reflection. In the case of collinear antiferromagnetic interlayer ordering, the "magnetic" maxima on the reflectivity angular curve are formed practically only by radiation with this rotated polarization. The first experiment on Mössbauer reflectivity with a selection of the rotated polarization discovers the predicted peak near the critical angle. The measurement of the rotated π→σ polarization component in Mössbauer reflectivity spectra excludes the interference with non-resonant electronic scattering and simplifies the spectrum shape near the critical angle allowing for an improved data interpretation in the case of poorly resolved spectra. It is shown that the selected component of Mössbauer reflectivity with rotated polarization is characterized by enhanced surface sensitivity, determined by the "squared standing waves" depth dependence. Therefore, the new approach has interesting perspectives for investigations of surfaces, ultrathin layers and multilayers having complicated magnetic structures.

Keywords: X-ray reflectivity; Mössbauer spectroscopy; magnetic multilayers; standing waves

---

1. Introduction

Interaction of light with magnetized media is characterized by specific polarization dependences. That is true as well for X-ray radiation. Modern sources of synchrotron radiation produce X-rays of any desired polarization and polarization-dependent absorption or scattering near the X-ray absorption edges (XMCD—X-ray magnetic circular dichroism, XMLD—X-ray magnetic linear dichroism, XRMR—X-ray resonant magnetic reflectivity) [1–7] have become extremely effective methods of magnetic investigation. For non-resonant X-ray scattering, the polarization analysis has been applied for separation of the spin and orbital magnetic moments and magnetic structure investigations [8–13]. Polarization analysis has been used for the observation of the X-ray Faraday and Kerr effects with soft X-rays [14–22].

For Mössbauer radiation, the splitting of the nuclear levels by hyperfine interactions means simultaneously the energy separation of the absorbed or reemitted quanta by their polarization states. Theoretical description of the elliptical polarization for different hyperfine transitions was done long ago [23]. The polarization dependences of Mössbauer absorption and Faraday rotation in thick samples were theoretically developed and experimentally proved in the excellent paper of Blume and Kistner [24]. In conventional Mössbauer spectroscopy, the radioactive sources emit an un-polarized single line radiation and the polarization state of different absorption lines has not been of special interest. However, the polarization state of various absorption lines reveals itself in the ratio of their intensity. In this way, the spectrum shape characterizes the direction of the sample magnetization [25].

The nuclear resonance ("Mössbauer") experiments with synchrotron radiation have been started by using the specific way of the nuclear response registration: by measuring the time evolution of the delayed nuclear decay after prompt SR (synchrotron radiation) pulse [26,27]. In this time-domain approach, the hyperfine splitting of nuclear levels leads to the quantum beats in the nuclear decay, and the polarization state of various hyperfine components becomes essential: it determines the result of their interference. The waves with orthogonal polarizations do not interfere. To be more precise, the coherent addition of the waves with orthogonal polarizations does not give the interference term in the resulting intensity [28]. For example, when magnetization of the sample is parallel to the beam, the four hyperfine transitions excited by σ-polarized SR give just one quantum beat frequency in nuclear decay [29]. However, the measurements of the scattered radiation with polarization selection immediately show all possible frequencies of the quantum beats, because the linear polarization, resulting from coherent addition of the two circular polarization, rotates with the time delay. That was splendidly demonstrated in the papers of Siddons et al. [30,31].

In time-domain Mössbauer spectroscopy, the polarization analysis of the scattered radiation has been found to be very helpful for the separation of the delayed nuclear scattering from huge prompt electronic scattering at the initial moment of pulse excitation. The electronic scattering does not change the polarization state of the incident radiation whereas the scattering at hyperfine nuclear sublevels gives the $90°$-rotated polarization component. Therefore, the polarization selection of this rotated component supplies very efficient suppression of the non-resonant prompt response, allowing nuclear decay monitoring from the very short delay times (after ~1 ns of their excitation) [30,32,33].

Recent developments of the nuclear resonance beamlines make it possible now the energy-domain Mössbauer spectroscopy with SR. In particular, the nuclear $^{57}FeBO_3$ monochromator (synchrotron Mössbauer source—SMS) has been installed at the ID18 beamline of the European synchrotron (ESRF) and at the BL11XU beamline of SPring-8 [34–37]. The key point of the SMS is the pure nuclear (111) or (333) reflection (forbidden for electronic diffraction) of the iron borate $^{57}FeBO_3$ crystal, which provides a single-line purely π-polarized 14.4 keV radiation within the energy bandwidth of 8 neV. The crystal should be heated at a specific temperature close to the Néel point of 348.35 K. In comparison with the laboratory experiments, the application of the π-polarized beam results in new features of Mössbauer spectra measured with SMS in absorption or reflection geometry [38]. Use of a diamond phase plate in addition to the $^{57}FeBO_3$ monochromator at the BL11XU beamline gives the new possibilities to perform measurements in forward and grazing-incidence geometries with various (linear, circular, elliptical) polarization states of radiation [39].

Polarization analysis of the resonantly reflected radiation by magnetic multilyers has not been used before. For comparison in the nonresonant magnetic X-ray scattering the polarization analysis was effectively used for magnetic structure investigations. In polarized neutron reflectivity, the spin-flip analysis provides very valuable information. Therefore, we suppose that polarization analysis in Mössbauer reflectivity should be useful.

In this work, the first results demonstrating the peculiarities of the nuclear resonant reflectivity (Mössbauer reflectivity) with SMS supplemented by the selection of the component with rotated $\pi \rightarrow \sigma$ polarization are presented. We show the differences in the Mössbauer reflectivity angular dependencies and Mössbauer reflectivity spectra measured without and with selection of the $\pi \rightarrow \sigma$ component, and we explain new features of the $\pi \rightarrow \sigma$ reflectivity using the X-ray standing wave approach. The practical significance of this new development for the complicated spectra treatment or depth-resolved investigations is also discussed.

2. Theory

The amplitudes of the nuclear resonant scattering in the forward direction, including the change of the polarization $\nu \rightarrow \nu'$, in the case of the dipole nuclear resonant transitions and in the presence of hyperfine splitting of the nuclear levels have the following expression [40,41]:

$$f_j^{nucl}(\omega, \nu \rightarrow \nu') = -\frac{1}{2\lambda}\,\sigma_{res}\frac{2L+1}{2I_e+1}\,f_j^{LM}\sum_{m_e,m_g}\frac{\frac{\Gamma_j}{2}\,|\langle I_g m_g L \Delta m | I_e m_e \rangle|^2}{\hbar\omega - E_{jR}(m_e, m_g) + \frac{i\Gamma_j}{2}}\left[\vec{h}_{j\Delta m} \circ \vec{h}^{\,*}_{j\Delta m}\right]_{\nu \rightarrow \nu'} \tag{1}$$

where $\hbar\omega$ is the photon energy, λ is the corresponding radiation wavelength. For 14.4 kev M1 transition in ^{57}Fe $L = 1$, $I_e = 3/2$, $I_g = 1/2$, m_e, m_g are the magnetic quantum numbers, $\langle I_g m_g L \Delta m | I_e m_e \rangle$ are the Clebsch–Gordan coefficients, $\sigma_{res} = 2.56 \times 10^{-4}$ nm^2 is the resonant cross-section, $\lambda = 0.086$ nm, j numerate the kinds of the hyperfine splitting (i.e., different multiplets in Mössbauer spectrum), f_j^{LM} is the Lamb–Mössbauer factor, $\hat{h}_{\Delta m}$ in (1) are the spherical unit vectors in the hyperfine field principal axis $\vec{h}_x$, $\vec{h}_y$, $\vec{h}_z$:

$$\vec{h}_{\pm 1} = \mp i\,\frac{\vec{h}_x \pm i\vec{h}_y}{\sqrt{2}},\quad \vec{h}_0 = i\vec{h}_z, \tag{2}$$

and the sign $\circ$ designates the outer product of these spherical unit vectors.

Considering grazing incidence and specifying the orientation of the hyperfine magnetic field $\mathbf{B}_{hf}$ by polar β and azimuth γ angles, the angular dependences of the nuclear resonant scattering amplitude for different hyperfine transitions $\Delta m = m_e - m_g = \pm 1$, 0 can be presented as matrices in $\sigma-$, π-polarization basis vectors:

$$f_{\Delta m=0}^{nucl,\perp} \propto \begin{pmatrix} \sin^2\beta\,\cos^2\gamma & -\sin\beta\,\cos\beta\,\cos\gamma \\ -\sin\beta\,\cos\beta\,\cos\gamma & \cos^2\beta \end{pmatrix} \tag{3}$$

$$f_{\Delta m=\pm 1}^{nucl,\perp} \propto \frac{1}{2}\begin{pmatrix} \sin^2\gamma + \cos^2\gamma\,\cos^2\beta & (\cos\beta\,\cos\gamma \mp i\sin\gamma)\sin\beta \\ (\cos\beta\,\cos\gamma \pm i\sin\gamma)\sin\beta & \sin^2\beta \end{pmatrix} \tag{4}$$

(we determine β relative the sample normal and choose $\gamma = 0°$ for the direction in surface plane perpendicular to the beam) The non-diagonal matrix elements of the nuclear resonant scattering mean the appearance of the 90° rotated polarization components in the scattered radiation.

For magnetic dipole (M1) nuclear transition (as it takes place for 14.4 kev transition in ^{57}Fe), the matrices in (3), (4) should be considered for the magnetic field of radiation. Therefore, the vector-column of the magnetic field of radiation for the π-polarized incident radiation from SMS is represented as $\begin{pmatrix} 1 \\ 0 \end{pmatrix}$, and the first columns in (3), (4) describe the angular dependences and polarization properties of the amplitudes of the nuclear resonant scattering $f_{\Delta m}^{nucl}$ in our case. It follows from (3), (4) that for $\Delta m = 0$ transitions the rotated $\pi \rightarrow \sigma$ polarization component appears in the scattering intensity only if $\mathbf{B}_{hf}$ has a non-zero projection on the normal to the surface. For $\Delta m = \pm 1$ transitions the rotated $\pi \rightarrow \sigma$ polarization component is created if $\mathbf{B}_{hf}$ lies in the surface plane ($\beta = 90°$, but not for $\gamma = 0°$ and maximal for $\beta = 90°$, $\gamma = 90°$). Later we consider such planar magnetic structures, typical for thin films.

Mössbauer absorption spectra are determined by the imaginary part of the scattering amplitude (1) according to the optical theorem: $\sigma(\omega) = 2\lambda \sum_j Im f_j^{nucl}(\omega)$ ($\sigma(\omega)$ is the absorption cross section).

Mössbauer reflectivity spectra are not similar to the Mössbauer absorption spectra, but are distorted by the interference with the electronic scattering, and their shape strongly depends on the grazing angle. In the simplest case of semi-infinite mirror Mössbauer reflectivity spectra are calculated with the Fresnel formula, in which the refractive index is simply connected with the scattering amplitudes by electrons f^{el} and nuclei f^{nucl} (diagonal components in (1)): $n = 1 + \frac{\lambda^2}{2\pi}\rho\left(f^{el} + f^{nucl}\right)$ (ρ is the density of scatters) [42]. In the case of multilyers the multiple interference of waves reflected by all boundaries and subjected to the polarization transformation essentially complicates the theory. The total algorithm of the Mössbauer reflectivity calculations based on the 4×4-propagation matrices is rather lengthy and is not presented here. It has been described in many papers [43–45]. In the kinematical approximation, which is applicable at the larger grazing angles than the critical angle, the shape of the Mössbauer reflectivity spectra can be qualitatively described by $\left|\sum_j f_j^{nucl}(\omega)\right|^2$ for thin single layer and we can use (1) for evaluation of the ratio of lines in Mössbauer reflectivity spectra. For several layers or periodic structures, the phase shifts for the scattered waves should be taken into account. For instance, for the structure with antiferromagnetic interlayer coupling between ^{57}Fe layers the angular and polarization dependencies for the spectrum lines are described by other polarization matrices than (3), (4) (see Ref. [38]).

We start with the model calculations of the Mössbauer reflectivity angular curves and Mössbauer reflectivity spectra at several grazing angles using our computer code RESPC (available from the ESRF website [46]). The essential point is that the calculations have been performed with selection of the reflected radiation by the polarization state and the result is shown in Figure 1. The $\pi\rightarrow\pi$ reflectivity is drawn by thicker blue lines, $\pi\rightarrow\sigma$ reflectivity is drawn by thinner red lines with symbols.

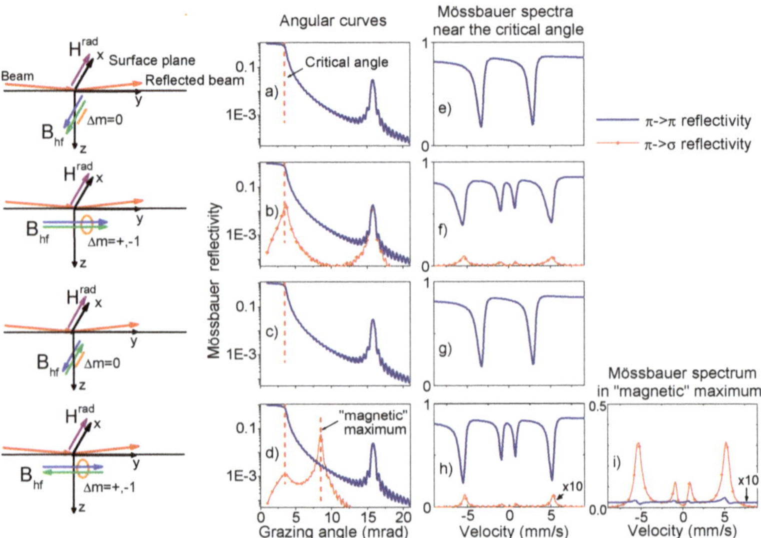

Figure 1. Model calculation of the angular dependencies of the Mössbauer reflectivity (**a–d**) and Mössbauer reflectivity spectra at the critical angle (**e–h**) and in the "magnetic" maximum (**i**) for π-polarized SMS radiation and for different cases of the ferromagnetic and antiferromagnetic coupling between adjacent ^{57}Fe layers, schematically drawn on the left.

In the energy-domain, the angular dependence of the Mössbauer reflectivity can be calculated either for a selected energy in the resonant spectrum range, or as the integral over the entire Mössbauer reflectivity spectra at each grazing angle θ. Figure 1a–d show the results obtained by the integrated mode, this mode corresponds to the experimental procedure with SMS. Calculations have been done for the $[^{57}\text{Fe}(0.8\ \text{nm})/\text{Cr}(2\ \text{nm})]_{30}$ multilayer, in ^{57}Fe layers we assume the presence of the hyperfine magnetic field of $B_{hf} = 33$ T with $\Delta B_{hf} = 1$ T distribution. We have considered the two magnetization directions relative to the radiation beam and the two types of the interlayer coupling between adjacent ^{57}Fe layers (ferromagnetic or antiferromagnetic), schematically shown by the blue and green arrows in the left column of Figure 1. Note that 14.4 keV Mössbauer transition is of magnetic dipole M1 type, so the magnetic field of radiation H^{rad} interacts with ^{57}Fe nuclei. In the case of the π-polarized radiation from SMS the radiation field vector H^{rad} lies in the sample surface. The hyperfine nuclear transitions ($\Delta m = 0$, $\Delta m = \pm 1$) allowed for π-polarized radiation are also indicated by orange lines in these sketches.

The results of these simple calculations show some unexpected features, not reported in previous experimental studies. The shapes of the angular curves of the Mössbauer $\pi \rightarrow \pi$ and $\pi \rightarrow \sigma$ reflectivity are found very different. For Mössbauer $\pi \rightarrow \pi$ reflectivity the angular dependences $I^{\pi \rightarrow \pi}(\theta)$ behave as usual X-ray reflectivity (they tend to 1 when θ). This behavior follows directly from usual Fresnel law, and it had been observed in the first paper devoted to Mössbauer reflectivity [42].

On the contrary, the angular dependencies $I^{\pi \rightarrow \sigma}(\theta)$ of the Mössbauer $\pi \rightarrow \sigma$ reflectivity have a sharp peak at the critical angle of the total external reflection and approach zero when $\theta \rightarrow 0$, as it seen in Figure 1b,d.

For the antiferromagnetic interlayer coupling, the magnetic period of the structure is twice larger than the chemical period. In this case, the additional Bragg peak ("magnetic" maximum) appears in the Mössbauer reflectivity angular dependence (see Figure 1d). Our calculations show that only the radiation with rotated $\pi \rightarrow \sigma$ polarization contributes to this peak (provided that there are no asymmetrically canted B_{hf} in the adjacent ^{57}Fe layers). Accordingly, the Mössbauer spectrum at the magnetic maximum is determined practically only by $\pi \rightarrow \sigma$ scattering (see Figure 1i). The rotated polarization component appears in the reflected signal only when the hyperfine field B_{hf} has a finite projection on the beam direction, i.e., when the excitation of the resonant $\Delta m = \pm 1$ transitions leads to the reemission of radiation with some part of the circular polarization. Accordingly, the Mössbauer reflectivity spectra with rotated polarization show not six but only four lines corresponding to the $\Delta m = \pm 1$ transitions as it takes place in Figure 1f,h,i. In the other cases, there is no reflectivity with rotated polarization at all.

Note that the discovered behavior of the angular dependencies of the Mössbauer $\pi \rightarrow \sigma$ reflectivity resembles much the angular dependence of the delayed nuclear resonant reflectivity measured in time-domain experiments [47]. For the delayed reflectivity, the vanishing of the delayed photons at zero grazing angle also can be explained by the Fresnel law: when $\theta \rightarrow 0$ the reflectivity for all energies in Mössbauer spectrum approaches unity, entirely losing its energy dependence. Accordingly, the intensity of the delayed radiation, determined by the Fourier transform of the energy dependent reflectivity amplitude, approaches zero. The physical explanation of the peak on the delayed reflectivity curve has been suggested in [48,49], and it is based on the phenomena of the X-ray standing waves. It has been shown that these waves, created by the prompt SR pulse, are responsible for the excitation of resonant nuclei. The X-ray standing wave explanation works in the case of the Mössbauer $\pi \rightarrow \sigma$ reflectivity in the energy-domain as well.

The basis for such consideration is the expression, obtained in [49], for the reflectivity from an ultrathin layer r' in the case when this ultrathin layer is placed at some depth z in a multilayer:

$$r'(z) = T(z)\,T'(z)\,(1 + R^{below}(z))^2\,r \tag{5}$$

where

$$r = i\frac{\pi \chi}{\lambda \sin \theta} d \tag{6}$$

is the reflectivity from an separated ultrathin layer of thickness d, θ is the grazing angle,

$$\chi = \frac{\lambda^2}{\pi}\rho f \tag{7}$$

is the susceptibility of this layer, ρ is the volume density of the scattering centers and f the amplitude of scattering in forward direction. $R^{below}(z)$ corresponds to the reflectivity amplitude from the part of the multilayer below the ultrathin layer, and functions $T(z)\,T'(z)$ describe the transformations of the transmitted and outgoing waves during multiple reflections at all boundaries in the upper part of the multilayer:

$$T(z)T'(z) = e^{2i(\varphi_1+\varphi_2+...+\varphi_{j-1})} \times \frac{\left(1-r_1^2\right)\left(1-r_2^2\right)...\left(1-r_{j-1}^2\right)}{\left(1+r_1 R_2\, e^{2i\varphi_1}\right)^2\left(1+r_2 R_3\, e^{2i\varphi_2}\right)^2...\left(1+r_{j-1} R_j\, e^{2i\varphi_{j-1}}\right)^2} \tag{8}$$

In (8) r_j and R_j are the Fresnel and multiple reflectivity amplitudes respectively at each boundary above the ultrathin layer,

$$\varphi_j = \frac{2\pi}{\lambda} d_j \sqrt{\sin^2\theta + \chi_j}, \tag{9}$$

is the phase shift for the waves during their transmission through each layer j. It is easy to see that

$$T(z)\, T'(z)\, \left(1 + R^{below}(z)\right)^2 = E^2(z), \tag{10}$$

where $E(z)$ is the total amplitude of the radiation field at depth z and $|E(z)|^2$ is no other than a standing wave. Assuming that f, χ and r in (7), (6) are the matrices in the case of the anisotropic scattering like (3), (4), the following expression can be written for the off-diagonal component of the reflectivity (supposing sufficiently small) from the whole multilayer in the case of π-polarization of the incident radiation:

$$I^{\pi\to\sigma}(\theta,\omega) = \frac{\lambda^2}{\sin^2\theta}\left|\int \rho_{res}(z)\, f_{res}^{\pi\to\sigma}(z,\omega)\, E_\pi^2(\theta,z,\omega)\, dz\right|^2. \tag{11}$$

This component appears only due to the nuclear resonance contribution to the susceptibility and $f_{res}^{\pi\to\sigma}$ is determined by the angular dependencies in (3), (4). The accurate derivation of (11) is presented in Ref. [50], and its application to the X-ray resonant magnetic reflectivity has been demonstrated in Ref. [51].

The presented expression (11) has important consequences on the features of the Mössbauer reflectivity with rotated polarization. Firstly, it explains perfectly the peak near the critical angle on the angular curve for the rotated $\pi\to\sigma$ reflectivity, because it contains the full radiation field $E(\theta,z,\omega)$ in the 4th power ("squared standing wave"). It is well known that the angular dependence of the secondary radiation, excited by X-ray standing waves, is characterized by the peak at the total reflection angle [52–54]. Secondly, the secondary radiation emission depends on the normal standing waves $|E(\theta,z)|^2$. The $\left|E^2(\theta,z,\omega)\right|^2$ factor in (11) ("squared standing wave") supplies the enhanced depth contrast comparing with $|E(\theta,z,\omega)|^2$, which is illustrated by Figure 2. Thirdly, the expression (11) essentially simplifies and fastens the calculations of the reflectivity with rotated polarization from anisotropic multilayers. It means that if the dichroic contribution to the scattering is small, the standing waves $E(\theta,z,\omega)$ at depth z of the sample can be calculated using the ordinal scalar reflectivity theory, and the "dichroic" response $I^{\pi\to\sigma}(\theta,\omega)$ is calculated by simple summation the scattering amplitudes $f_{res}^{\pi\to\sigma}(z,\omega)$ from different depth with proper "weights" (the generalized kinematical approximation). If the nuclear resonance scattering is strong enough, the radiation field $E_\pi(\theta,z,\omega)$ varies across the resonant spectrum (as seen in Figure 3b); however, in most cases these variations of $E_\pi(\theta,z,\omega)$ as a function of ω can be neglected especially if we measure the NRR (Nuclear Resonance Refelctivity) curves by integrating over Mössbauer reflectivity spectra. In this case the calculations become extremely fast.

The enhanced depth selectivity of the Mössbauer $\pi\to\sigma$ reflectivity spectra compared with the ordinal Mössbauer reflectivity spectra is illustrated by Figure 3. In order to distinguish the thin top layer (1.5 nm ^{57}Fe) from the bottom one, we assumed for the top layer the hyperfine magnetic field B_{hf} = 28 T, and for the deeper ^{57}Fe layer (3 nm thickness) B_{hf} = 30 T. The calculated Mössbauer reflectivity spectra at several grazing angles near the critical one without and with $\pi\to\sigma$ polarization selection clearly show that the $\pi\to\sigma$ reflectivity spectra contain more intense contribution from the 1.5 nm top layer than the Mössbauer reflectivity spectra without polarization selection.

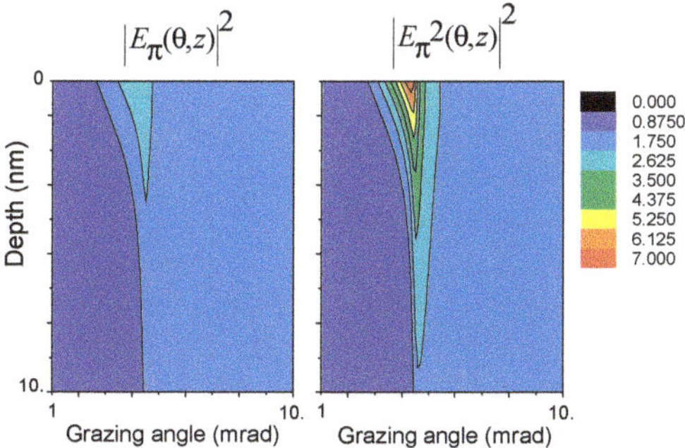

Figure 2. Comparison of the angular-depth dependence of the standing wave $|E_\pi(\theta, z, \omega)|^2$ (**left panel**) and squared standing wave $\left|E_\pi{}^2(\theta, z, \omega)\right|^2$ (**right panel**), calculated for an iron mirror neglecting the nuclear resonant contribution to the scattering.

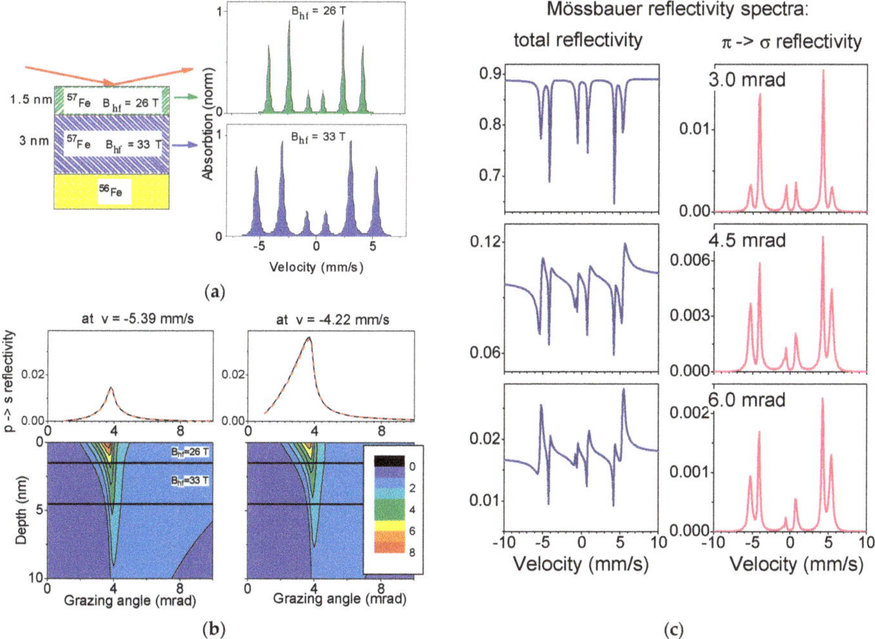

Figure 3. (**a**) The used in calculations model: two thin layers with slightly different B_{hf} (their resonant spectra are on the left), directed along the beam. (**b**, **bottom panels**) The angular-depth distribution of the squared standing waves $\left|E_\pi{}^2(\theta, z, \omega)\right|^2$, calculated for two indicated velocities (v) points in the resonant spectrum, the solid horizontal lines show the layers boundaries, (**b**, **upper panels**) the angular curves for $\pi \to \sigma$ reflectivity, corresponding to this energy points. The solid and dashed lines (nearly indistinguishable) show the results, obtained by Equation (1) and by the exact calculations, performed with the program REFSPC [46]. (**c**) The Mössbauer reflectivity spectra without the polarization selection (**left column**) and for $\pi \to \sigma$ reflectivity (**right column**), calculated for three indicated grazing angles in vicinity of the critical angle.

The validity of the generalized kinematical approximation (11) is also checked by calculations presented in Figure 3b. The angular curves in Figure 3b are calculated in two ways: by the formula (11) and by the exact calculations performed with the program pack REFSPC [46] (solid and dashed lines respectively). Both results coincide. However, in order to achieve this practically full coincidence, the nuclear resonant susceptibility was decreased by a factor of 5 relative to the value for the α-iron films fully enriched by ^{57}Fe isotope. Such decreased nuclear resonance susceptibility is typical for real thin films with lower enrichment and broad distribution of the hyperfine magnetic fields. Thereby, the direct comparison between the kinematical approximation (11) and the exact calculations shows the excellent agreement, provided that the resonant contribution to the susceptibility is not too large.

The next advantage of the measurements with the selection of polarization is illustrated by Figure 4. It shows that the selection of polarization leads to the decreasing number of lines in the Mössbaur reflectivity spectra. This can be very useful for the interpretation of the poorly resolved complicated spectra. The experimental Mössbaur reflectivity spectrum from Ref. [55], presented on the top of Figure 4, can be fitted by at least three different models with completely different field orientation and distribution of the hyperfine magnetic field. The calculated $\pi \rightarrow \sigma$ reflectivity spectra are quite different for these three cases. Therefore, the true picture of the hyperfine field distribution and orientation can be recovered if one performs the polarization analysis of the reflected radiation.

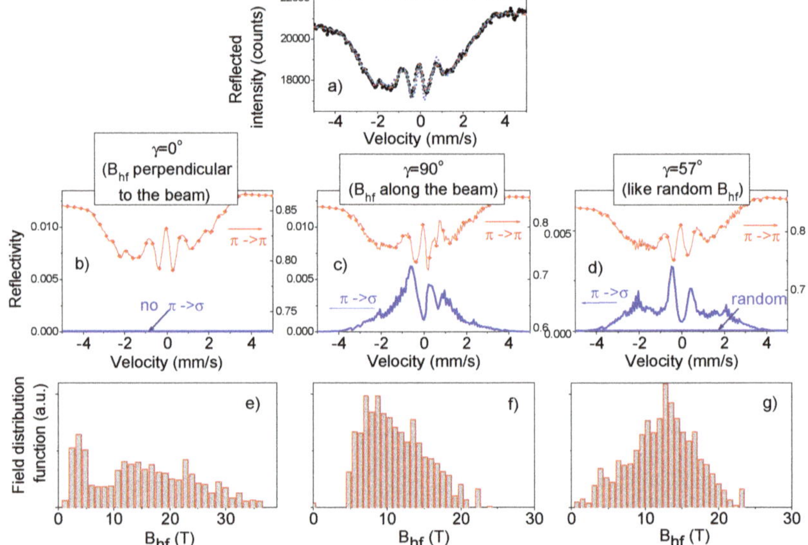

Figure 4. (**a**) Mössbauer reflectivity spectrum from Ref. [55], measured without polarization selection for the $[^{57}\text{Fe}(0.12 \text{ nm})/\text{Cr}(1.05 \text{ nm})] \times 30$ sample in remanence. Symbols are the experimental data, three solid (practically undistinguishable) lines are the results of the fit with three different orientations of the hyperfine magnetic field B_{hf}: for B_{hf} in the surface plane and perpendicular to the beam (azimuthal angle $\gamma = 0°$), B_{hf} parallel to the beam ($\gamma = 90°$), and B_{hf} in the surface plane with $\gamma = 57°$ (for this angle the spectrum without polarization analysis is the same as in the case of the random in space orientations of B_{hf}). (**b–d**) Theoretical Mössbauer reflectivity spectra for these models with the selection of polarization: solid red lines with symbols are the results for the reflectivity with nonrotated ($\pi \rightarrow \pi$) polarization, solid blue lines are those for the rotated ($\pi \rightarrow \sigma$) polarization component in the reflectivity. (**e–g**) The corresponding different field distributions P(B_{hf}) for these three cases, giving the same result of the fit to the experimental spectrum shown in (**a**).

3. Experimental Results and Discussion

The experiment has been performed at the Nuclear Resonance beamline [56] ID18 of the ESRF. Figure 5 shows the experimental setup. The storage ring was operated in multi-bunch mode with a nominal current of 200 mA. The energy bandwidth of radiation was first reduced down to 2.1 eV by the high-heat-load monochromator [57] adjusted to the 14.4125 keV energy of the nuclear resonant transition of the ^{57}Fe isotope. Then X-rays were collimated by the compound refractive lenses down to the angular divergence of a few µrad. The high-resolution monochromator decreases the energy bandwidth of the beam further to ~15 meV. Final monochromatization down to the energy bandwidth of ~8 neV was achieved with the pure nuclear (111) reflection of the ^{57}FeBO$_3$ crystal, and the sweep through the energy range of a Mössbauer resonant spectrum (about ±0.5 µeV) was achieved using the Doppler velocity scan. Radiation from the SMS was focused vertically using the compound refractive lenses down to the beam spot of 50 µm. The intensity of the X-ray beam incident on the sample was about 10^4 photons/s. More details on the design of the SMS can be found in Ref [37].

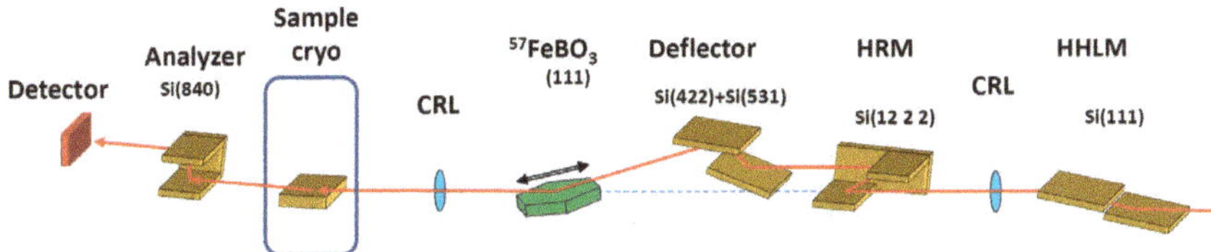

Figure 5. Experimental set-up for measurements of the Mössbauer reflectivity with the selection of radiation with the rotated ($\pi\rightarrow\sigma$) polarization. HHLM—high-heat-load monochromator; CRL—compound refractive lenses, HRM—high-resolution monochromator.

The selection of the radiation with the rotated ($\pi\rightarrow\sigma$) polarization was performed by silicon channel-cut crystal with two asymmetric Si(840) reflections with the Bragg angle $\theta_B = 45.1°$. According to theoretical calculations, the channel-cut should suppress the π-polarized radiation by about five orders of magnitude, while transmitting about 70% of the σ-polarized radiation with the angular acceptance of about 2.6 arc sec (FWHM).

We studied several [^{57}Fe/Cr]$_{30}$ samples. The samples were grown with the Katun-C molecular beam epitaxy facility, equipped by 5 thermal evaporators at the Institute of Metal Physics (Ekaterinburg, Russia). The growth was performed under the UHV (Ultra-high vacuum) regime (5×10^{-10} mbar). During the buffer layer deposition, the Al$_2$O$_3$ substrate temperature was gradually decreased from 300 °C down to 180 °C. The typical deposition rate of Cr and ^{57}Fe layers was about of 0.15 nm/min. The details on the preparations and characterization of the samples can be found elsewhere [38]. The samples were mounted in the cassette holder of the He-exchange gas superconducting cryo-magnetic system. The studies were performed at helium temperature (4.0 K).

The angular scan with the polarization analyzer, performed for all used samples in order to catch the $\pi\rightarrow\sigma$ reflected radiation, demonstrated the extremely broadened divergence of the reflected signal as it is shown in Figure 6 for one of the sample. In essence, the samples were composed of many nano-crystallites, essentially misoriented relative to each other. Consequently, the radiation reflected by the sample consisted of many spikes, spreading over the angular range of about 200 arc sec as seen in Figure 6. For such a divergent reflected beam, the throughput of the Si(840) channel-cut analyzer for the σ-polarized radiation was only about 4%. Therefore, the measurements of the reflectivity with the rotated $\pi\rightarrow\sigma$ polarization were essentially complicated by the nano-islands structure of the studied samples. Later we get the GISAXS (Grazing-incidence small-angle X-ray scattering) pattern from this sample which shows the cluster-layered structure of the whole film [58].

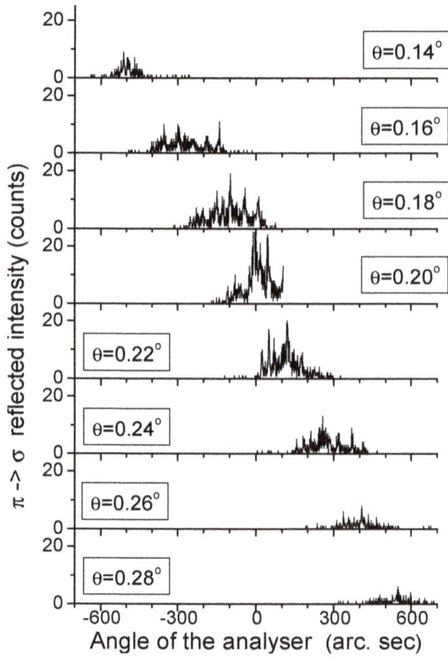

Figure 6. Angular distributions of the radiation reflected with $\pi{\rightarrow}\sigma$ rotation of the polarization from the $[^{57}Fe(0.8\ nm)/Cr(1.05\ nm)]_{30}$ sample measured with the Si(840) channel-cut analyzer for various grazing angles θ.

With the resulting very small count rate, the measurements at the weak "magnetic" maxima for the samples with antiferromagnetic interlayer coupling were not feasible, and we decided to concentrate on measurements near the critical angle. In order to ensure the ferromagnetic alignment of the magnetization in the ^{57}Fe layers and increase the Mössbauer dichroic reflectivity, the external magnetic field of 5 T was applied along the beam direction.

In order to increase the count rate, the temperature of the $^{57}FeBO^3$ crystal for this particular measurement was lowered by few degrees. This sacrifices the energy resolution of the SMS to ~230 neV (~5 mm/s), which is not important for the angular dependency measurements, but increases the intensity of the SMS by an order of magnitude [37]. The nuclear resonance reflectivity with the rotated polarization shown in Figure 7 was obtained for each grazing angle θ by integrating over the angular distributions measured with the Si(840) analyzer and drawn in Figure 6.

The predicted peak near the critical angle for the Mössbauer $\pi \rightarrow \sigma$ reflectivity is clearly seen in Figure 7. The origin of this peak is already explained in the previous section. The angular dependence of the Mössbauer reflectivity measured without the polarization selection shows the plateau below the critical angle. Note that the signal values for two curves in Figure 7 are not possible to compare because they are measured with different incident intensities from $^{57}FeBO_3$ nuclear monochromator.

Mössbauer $\pi \rightarrow \sigma$ reflectivity spectra were measured at the two grazing angels, $\theta = 0.19°$ and $\theta = 0.22°$, in vicinity of the critical angle. They are shown in Figure 8. In spite of the limited statistics, caused by the essential loss of the intensity due to the nano-islands structure of the samples and narrow angular reception by ideal Si crystal-analyzer, we can recognize in these data some important features. The resonant lines are presented in these spectra as peaks, as predicted by the theory (see Figure 1f,h,g). The spectrum measured without polarization analysis is also presented in Figure 8 for comparison. The lines in this spectrum have a dispersive-like shape and they are essentially broadened, as it should be at the angles near the critical angle [46]. Note, that in spite of the higher statistical accuracy of the spectrum measured without the polarization analysis, its smeared shape makes the correct interpretation more difficult. Thus, the spectra of reflectivity with rotated polarization should be more suited for the right data analysis, provided that better statistics will be reached by optimization of the experiment.

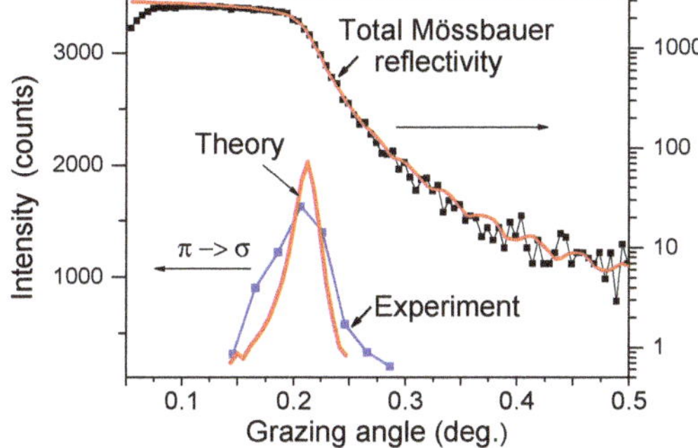

Figure 7. Angular dependencies of the Mössbauer reflectivity measured without selection of polarization and with the selection of the rotated $\pi \to \sigma$ component in reflectivity, obtained as the integral over scans, presented in Figure 6. The lines with symbols show the experimental data. The red solid lines show the results of theoretical calculations.

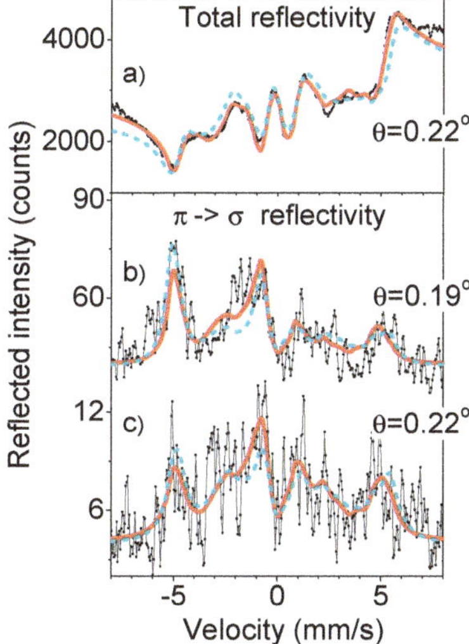

Figure 8. Mössbauer reflectivity spectra measured from $[^{57}\text{Fe}(0.8 \text{ nm})/\text{Cr}(1.05 \text{ nm})]_{30}$ multilayer without polarization analysis at the critical angle (**a**) and with selection of the rotated polarization component at the two angles in the vicinity of the critical angle (**b,c**). The solid lines with symbols are the experimental data. The red sold and green dashed lines are the fits discussed in the text.

The fit of the $\pi \to \sigma$ reflectivity spectra is not really reliable due to insufficient statistics, we merely wanted to show what treatment should be performed with better experimental data and demonstrate the surface sensitivity of the method. So, the shapes of the spectra in Figure 8 can be more or less reproduced by the fit performed with the REFSPC program package [46]. All three spectra are treated within the same model. We have used seven hyperfine fields $B_{hf}^{(i)}$, i = 1, 2, ... , 7, and the fitted parameters and distribution probabilities of the fields $P(B_{hf})$ are shown in Figure 9a.

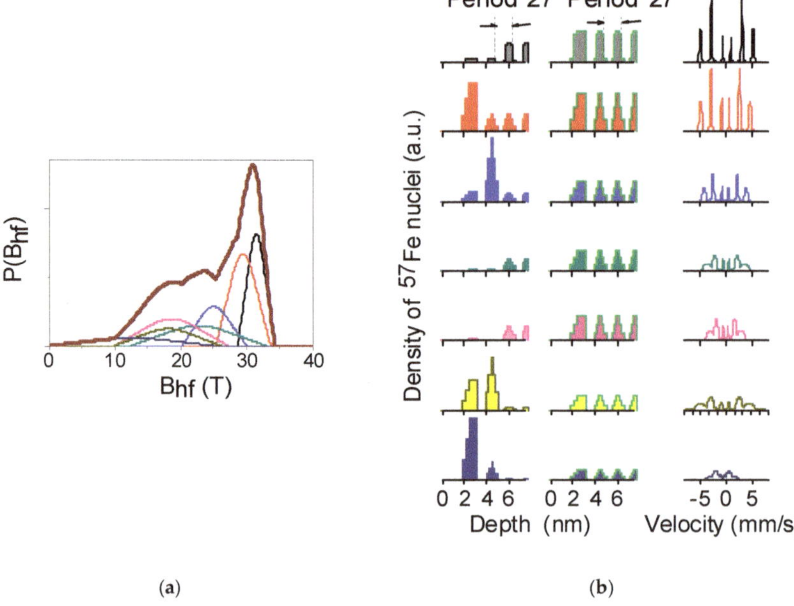

Figure 9. (**a**) Hyperfine field distribution, obtained by the fit of the spectra in Figure 8. (**b**) Depth distribution of the each kind of $B_{hf}^{(i)}$ (i = 1, 2, ... , 7) in several top ^{57}Fe layers (the period, repeating 27 times in the deeper part of the multilayer, is marked, the last period contacted with the buffer layer is also different from the main periodic part but is not shown here), obtained by the fit (**left column**); the initially suggested homogeneous depth distribution of $B_{hf}^{(i)}$ in all ^{57}Fe layers (**middle column**), the calculated reflectivity spectra for this distribution are presented in Figure 8 by the dashed green lines. In the right column, the resonant spectra corresponding to every $B_{hf}^{(i)}$ are shown.

The depth profiles obtained for each field $B_{hf}^{(i)}$ are presented in the left column of Figure 9b. We can see that the hyperfine fields in the top two ^{57}Fe layers are significantly different from the fields in the deeper layers: on average, in the top layers the hyperfine splitting is smaller, and the fraction of the low field contributions is larger. Assuming the same field distribution in the top layers as in the whole periodic part (shown in Figure 9b, middle column), we obtain the theoretical spectra, presented by dashed green lines in Figure 8. They reveal worse correspondence with the experimental spectra. These calculations confirm the enhanced depth selectivity of the spectra of reflectivity with rotated $\pi{\to}\sigma$ polarization.

Other unexpected features of the measured spectra were noticed. We were surprised by the asymmetry and by the difference in shapes of the $\pi \to \sigma$ reflectivity spectra measured at the two close angels: θ = 0.19° and θ = 0.22° (compare the spectra in Figure 8b,c). At the beginning we supposed that these effects are caused by a small contribution of the reflectivity with the non-rotated polarization. But the model calculations did not confirm such assumption. Finally, the effect was explained by the refraction effect.

For illustration of this effect we use a simple model. Calculations have been performed for the 10 nm ^{57}Fe layer where two hyperfine fields B_{hf} = 28 T and 30 T take place, the resonant spectra for these two fields are drown in Figure 10b. The angular dependencies of the Mössbauer $\pi \to \sigma$ reflectivity, calculated for the photon energies corresponding to two outer lines in these spectra reveal the different angular positions of the critical angle peaks which is clearly seen in Figure 10c. This shift of the peak positions explains the asymmetry of the rotated $\pi{\to}\sigma$ polarization component in Mössbauer reflectivity spectra.

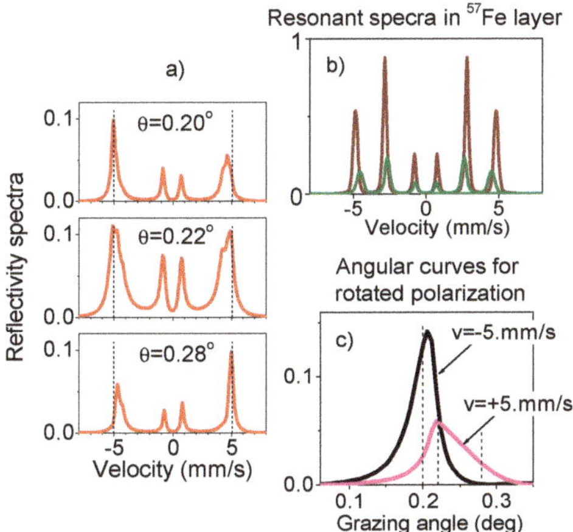

Figure 10. (a) Model calculations of the rotated $\pi{\rightarrow}\sigma$ polarization component in Mössbauer reflectivity spectra. Dashed vertical lines mark the energies, for which the angular dependencies of Mössbauer reflectivity in (c) are calculated. (b) The resonant spectra for the B_{hf} = 30 T and 28 T which we assume in the ⁵⁷Fe layer. (c) The angular dependencies of the Mössbauer reflectivity with the rotated $\pi{\rightarrow}\sigma$ polarization, calculated for the 1st (v = −5 mm/s) and 6th lines (v = +5 mm/s) of the resonant spectrum corresponding to B_{hf} = 30 T. The dashed vertical lines mark the angles, for which the spectra in (a) are calculated.

The position of the critical angle is determined by the real part of the susceptibility χ of the layer (or the refraction index $n = 1 + \chi/2$). For nuclear resonance scattering, this includes the electronic scattering part and the nuclear resonant contribution. The real part of the nuclear resonant contribution has the opposite signs at two sides of the exact resonance. Therefore, the resonant lines of the narrower sextet with B_{hf} = 28 T have opposite influences on the refraction index and, accordingly, on the critical angle position for the radiation wavelengths corresponding to the 1st and 6th lines of the broader sextet. Because the maximum of reflectivity for these two lines occurs at slightly different angles, the relative intensities of these two lines are not the same (as they are supposed to be for the absorption case), and they are changed with a small variation of the grazing angle. This provides the observed asymmetry of the spectra depending on the small variation of the grazing angle observed in Figure 8. Note that the studied sample shows even more low-field $B_{hf}^{(i)}$ contributions. However, the effect of all smaller fields is, in principle, the same as revealed by the simplest model calculations.

Note that the same changes of the asymmetry of the Mössbauer reflectivity spectra were observed with this particular sample in measurements at the "magnetic" maximum [59]. Although those measurements were performed without the polarization analysis, the "magnetic" maximum is formed basically by the scattering with the rotated $\pi{\rightarrow}\sigma$ polarization (as shown by the model calculations in Figure 1d,i).

4. Summary

We have performed the first experiment on the nuclear resonant reflectivity in the energy domain with the polarization selection, with the two asymmetric Si(840) reflections from a silicon channel-cut crystal. The experiment has been performed using the synchrotron Mössbauer source, benefiting from the purely linear π-polarized radiation of this instrument. We have measured the angular dependence of the $\pi{\rightarrow}\sigma$ reflectivity as well as the Mössbauer $\pi{\rightarrow}\sigma$ reflectivity spectra at two angles in vicinity of the total external reflection.

We have observed a new effect not reported before. Namely, the nuclear resonance (Mössbauer) $\pi{\rightarrow}\sigma$ reflectivity in the energy domain is characterized by a peak near critical angle of the total external reflection (instead of the Fresnel plateau) similar to that observed for the nuclear resonant reflectivity in the time domain. The effect is explained by the influence of the X-ray standing waves on the nuclear resonant scattering.

Our first experimental realization of the new method shows that, for the polarization analysis in reflectivity experiments, a thorough choice of the right way of polarization selection is needed. The reason is that the reflectivity from cluster-layered films has essentially broadened angular divergence due to a surface roughness and other imperfections. The channel-cut Si (840) reflection, effectively used in previous polarization experiments in forward scattering or for crystal diffraction [32,33] has angular acceptance that is too small and we lost most of the reflected radiation. For this reason, the quality of the measured Mössbauer $\pi \rightarrow \sigma$ reflectivity spectra was not good enough for quantitative interpretation and our fit just demonstrates the future possibilities. It is clear that other polarization analyzers should be tested (e.g., Be (006) reflection, LiF (622) reflection, Ge (664) reflection, a pyrolytic graphite (222) reflection or some others for 14.4 kev radiation etc.) in order to enlarge the registered signal. Note that in the soft X-ray region a rather unusual way of polarization selection is used (an electron time-of-flight polarization analyzer) [60]. In [61] an annealed silicon crystal was used in order to enlarge the angular acceptance of this monochromator/analyzer. We believe that the right choice of the polarization analyzer allows one to measure Mössbauer $\pi \rightarrow \sigma$ reflectivity with essentially better statistics and to use and demonstrate all theoretically described advantages of the new method of registration.

We have shown (at least by model calculations) that the polarization analysis in Mossbauer spectroscopy simplifies the Mössbauer spectra, restricting the number of lines. It should be helpful for their interpretation in cases of complicated hyperfine interactions (typical for Mössbauer nuclei at a surface, in nano-objects and in thin layers). Moreover, in the reflectivity geometry the spectra lost the dispersive-like features caused by the interference of nuclear and electronic scattering. The enhanced surface sensitivity of the reflectivity signal with rotated $\pi \rightarrow \sigma$ polarization is also predicted, which follows from the "squared-standing-wave" depth dependence of the scattered radiation.

Concluding, we believe that the described advantages of Mössbauer spectroscopy with polarization selection will help to resolve most important scientific cases in surface science, with new physics and high data quality, and will provide us with a more reliable interpretation of the magnetic ordering in multilayer films.

Author Contributions: Conceptualization, M.A.; methodology, A.C., R.R.; investigation, M.A., R.B., T.K., A.C.; software, M.A.; data treatment, R.B., T.K.

Funding: This work was supported by the Russian Foundation of the Basic Research through the Grants No. 15-02-1502 and 16-02-00887-a. R.B. acknowledges a personal grant from the Foundation for the Advancement of Theoretical Physics and Mathematics "BASIS" (grant 18-2-6-22-1).

Acknowledgments: The authors are thankful to Yu. A. Babanov, D. A. Ponomarev and M. A. Milyaev (all IMP UB RAS) for the assigned samples.

Conflicts of Interest: The authors declare no conflict of interest.

References

1. Schütz, G.; Knülle, V.; Ebert, H. X-ray magnetic circular dichroism and its relation to local magnetic structure. In *Resonant Anomalous X-Ray Scattering*; Materlik, G., Sparks, C.J., Fischer, K., Eds.; Elsevier Science: New York, NY, USA, 1994.

2. Stöhr, J. X-ray magnetic circular dichroism spectroscopy of transition metal thin films. *J. Electron Spectrosc. Relat. Phenom.* **1995**, *75*, 253–272. [CrossRef]

3. Kortright, J.B.; Awschalom, D.D.; Stöhr, J.; Bader, S.D.; Idzerda, Y.U.; Parkin, S.S.P.; Schuller, I.K.; Siegmann, H.-C. Research frontiers in magnetic materials at soft X-ray synchrotron radiation facilities. *J. Magn. Magn. Mater.* **1999**, *207*, 7–44. [CrossRef]

4. Van der Laan, G. Applications of soft X-ray magnetic dichroism. *J. Phys. Conf. Ser.* **2013**, *430*, 012127. [CrossRef]

5. Sorg, C.; Scherz, A.; Baberschke, K.; Wende, H.; Wilhelm, F.; Rogalev, A.; Chadov, S.; Minár, J.; Ebert, H. Detailed fine structure of x-ray magnetic circular dichroism spectra: Systematics for heavy rare-earth magnets. *Phys. Rev. B* **2007**, *75*, 064428. [CrossRef]
6. Sève, L.; Jaouen, N.; Tonnerre, J.M.; Raoux, D.; Bartolomé, F.; Arend, M.; Felsch, W.; Rogalev, A.; Goulon, J.; Gautier, C.; et al. Profile of the induced 5d magnetic moments in Ce/Fe and La/Fe multilayers probed by x-ray magnetic-resonant scattering. *Phys. Rev. B* **1999**, *60*, 9662–9674. [CrossRef]
7. Ishimatsu, N.; Hashizume, H.; Hamada, S.; Hosoito, N.; Nelson, C.S.; Venkataraman, C.T.; Srajer, G.; Lang, J.C. Magnetic structure of Fe/Gd multilayers determined by resonant X-ray magnetic scattering. *Phys. Rev. B* **1999**, *60*, 9596–9606. [CrossRef]
8. Blume, M.; Gibbs, D. Polariztion dependence of magnetic x-ray scattering. *Phys. Rev. B* **1988**, *37*, 1779–1789. [CrossRef]
9. Moncton, D.E.; Gibbs, D.; Bohr, J. Magnetic x-ray scattering with synchrotron radiation. *Nucl. Instrum. Methods Phys. Res. A* **1986**, *246*, 839–844. [CrossRef]
10. McWhan, D.B.; Vettier, C.; Isaacs, E.D.; Ice, G.E.; Siddons, D.P.; Hastings, J.B.; Peters, C.; Vogt, O. Magnetic x-ray-scattering study of uranium arsenide. *Phys. Rev. B* **1990**, *42*, 6007–6017. [CrossRef]
11. Bohr, J. Magnetic x-ray scattering: A new tool for magnetic structure investigations. *J. Magn. Magn. Mater.* **1990**, *83*, 530–534. [CrossRef]
12. Langridge, S.; Lander, G.H.; Bernhoeft, N.; Stunault, A.; Vettier, C.; Grübel, G.; Sutter, C.; de Bergevin, F.; Nuttall, W.J.; Stirling, W.G.; et al. Separation of the spin and orbital moments in antiferromagnetic UAs. *Phys. Rev. B* **1997**, *55*, 6392–6398. [CrossRef]
13. Sasaki, Y.; Okube, M.; Sasaki, S. Resonant and non-resonant magnetic scatterings with circularly polarized X-rays: Magnetic scattering factor and electron density of gadolinium iron garnet. *Acta Cryst. A* **2017**, *73*, 257–270. [CrossRef] [PubMed]
14. Siddons, D.P.; Hart, M.; Amemiya, Y.; Hastings, J.B. X-Ray Optical Activity and the Faraday Effect in Cobalt and Its Compounds. *Phys. Rev. Lett.* **1990**, *64*, 1967–1970. [CrossRef] [PubMed]
15. Kortright, J.; Rice, M. Soft X-ray magneto-optic Kerr rotation and element-specific hysteresis measurement. *Rev. Sci. Instrum.* **1996**, *67*, 3353. [CrossRef]
16. Collins, S.P. X-ray Faraday rotation and magnetic circular dichroism in an iron–platinum compound. *J. Phys. Condens. Matter.* **1999**, *11*, 1159–1175. [CrossRef]
17. Mertins, H.-C.; Schäfers, F.; Gaupp, A.; Le Cann, X.; Gudat, W. Faraday rotation at the 2p edges of Fe, Co, and Ni. *Phys. Rev. B* **2000**, *61*, R874–R877. [CrossRef]
18. Kortright, J.B.; Kim, S.-K. Resonant magneto-optical properties of Fe near its 2p levels: Measurement and applications. *Phys. Rev. B* **2000**, *62*, 12216–12228. [CrossRef]
19. Mertins, H.-C.; Oppeneer, P.M.; Kuneš, J.; Gaupp, A.; Abramsohn, D.; Schäfers, F. Observation of the X-Ray Magneto-Optical Voigt Effect. *Phys. Rev. Lett.* **2001**, *87*, 047401. [CrossRef]
20. Oppeneer, P. Magneto-optical Kerr spectra. *Handb. Magn. Mater.* **2001**, *13*, 229–422.
21. Mertins, H.C.; Valencia, S.; Abramsohn, D.; Gaupp, A.; Gudat, W.; Oppeneer, P.M. X-ray Kerr rotation and ellipticity spectra at the 2 p edges of Fe, Co, and Ni. *Phys. Rev. B* **2004**, *69*, 064407. [CrossRef]
22. Mertins, H.-C.; Valencia, S.; Gaupp, A.; Gudat, W.; Oppeneer, P.M.; Schneider, C.M. Magneto-optical polarization spectroscopy with soft X-rays. *Appl. Phys. A* **2005**, *80*, 1011–1020. [CrossRef]
23. Frauenfelder, H.; Nagle, D.E.; Taylor, R.D.; Cochran, D.R.; Visscher, W.M. Elliptical Polarization of Fe^{57} Gamma Rays. *Phys. Rev.* **1961**, *126*, 1065–1075. [CrossRef]
24. Blume, M.; Kistner, O.C. Resonant absorption in the presence of Faraday rotation. *Phys. Rev.* **1968**, *171*, 417–425. [CrossRef]
25. Housley, R.M.; Grant, R.W.; Gonser, U. Coherence and Polarization Effects in Mössbauer Absorption by Single Crystals. *Phys. Rev.* **1979**, *178*, 514–522. [CrossRef]
26. Gerdau, E.; Rüffer, R.; Winkler, H.; Tolksdorf, W.; Klages, C.P.; Hannon, J.P. Nuclear Bragg diffraction of synchrotron radiation in Yttrium Iron Garnet. *Phys. Rev. Lett.* **1985**, *54*, 835–838. [CrossRef]
27. Hastings, J.B.; Siddons, D.P.; van Bürck, U.; Hollatz, R.; Bergmann, U. Mössbauer spectroscopy using synchrotron radiation. *Phys. Rev. Lett.* **1991**, *66*, 770–773. [CrossRef] [PubMed]
28. Ramachandrah, G.S.; Ramaseshah, S. Crystal Optics. In *Handbuch der Physic*; Springer: Berlin, Germany, 1961; pp. 1–127.

29. Smirnov, G.V. General properties of nuclear resonant scattering. *Hyperfine Interact.* **1999**, *123*, 31–77. [CrossRef]

30. Siddons, D.P.; Bergmann, U.; Hastings, J.B. Time-dependent polarization in Mössbauer experiments with synchrotron radiation: Suppression of electronic scattering. *Phys. Rev. Lett.* **1993**, *70*, 359–362. [CrossRef]

31. Siddons, D.P.; Bergmann, U.; Hastings, J.B. Polarization effects in resonant nuclear scattering. *Hyperfine Interact.* **1999**, *123/124*, 681–719. [CrossRef]

32. Toellner, T.S.; Alp, E.E.; Sturhahn, W.; Mooney, T.M.; Zhang, X.; Ando, M.; Yoda, Y.; Kikuta, S. Polarizer/analyzer filter for nuclear resonant scattering of synchrotron radiation. *Appl. Phys. Lett.* **1995**, *67*, 1993–1995. [CrossRef]

33. L'abbé, C.; Coussement, R.; Odeurs, J.; Alp, E.E.; Sturhahn, W.; Toellner, T.S.; Johnson, C. Experimental demonstration of time-integrated synchrotron-radiation spectroscopy with crossed polarizer and analyzer. *Phys. Rev. B* **2000**, *61*, 4181–4185. [CrossRef]

34. Smirnov, G.V.; van Bürck, U.; Chumakov, A.I.; Baron, A.Q.R.; Rüffer, R. Synchrotron Mössbauer source. *Phys. Rev. B* **1997**, *55*, 5811–5815. [CrossRef]

35. Mitsui, T.; Seto, M.; Kikuta, S.; Hirao, N.; Ooishi, Y.; Takei, H.; Kobayashi, Y.; Kitao, S.; Higashitaniguchi, S.; Masuda, R. Generation and Application of Ultrahigh Monochromatic X-ray Using High-Quality ^{57}FeBO$_3$ Single Crystal. *Jpn. J. Appl. Phys.* **2007**, *46*, 821–825. [CrossRef]

36. Mitsui, T.; Hirao, N.; Ohishi, Y.; Masuda, R.; Nakamura, Y.; Enoki, H.; Sakaki, K.; Seto, M. Development of an energy-domain ^{57}Fe-Mossbauer spectrometer using synchrotron radiation and its application to ultrahigh-pressure studies with a diamond anvil cell. *J. Synchrotron Radiat.* **2009**, *16*, 723–729. [CrossRef] [PubMed]

37. Potapkin, V.; Chumakov, A.I.; Smirnov, G.V.; Celse, J.-P.; Rüffer, R.; McCammon, C.; Dubrovinsky, L. The ^{57}Fe Synchrotron Mössbauer Source at the ESRF. *J. Synchrotron Radiat.* **2012**, *19*, 559–569. [CrossRef] [PubMed]

38. Andreeva, M.A.; Baulin, R.B.; Chumakov, A.I.; Rüffer, R.; Smirnov, G.V.; Babanov, Y.A.; Devyaterikov, D.I.; Milyaev, M.A.; Ponomarev, D.A.; Romashev, L.N.; et al. Nuclear resonance reflectivity from [^{57}Fe/Cr]$_{30}$ multilayer with Synchrotron Mössbauer Source. *J. Synchrotron Radiat.* **2018**, *25*, 473–483. [CrossRef]

39. Mitsui, T.; Imai, Y.; Masuda, R.M.; Seto, M.; Mibu, K. ^{57}Fe polarization-dependent synchrotron Mössbauer spectroscopy using a diamond phase plate and an iron borate nuclear Bragg monochromator. *J. Synchrotron Radiat.* **2015**, *22*, 427–435. [CrossRef]

40. Trammell, G.T. Elastic Scattering at Resonance from bound nuclei. *Phys. Rev.* **1962**, *126*, 1045–1054. [CrossRef]

41. Andreeva, M.A.; Kuz'min, R.N. *Mössbauerovskaya Gamma Optika (in Russian)*; Moscow Universitet Publ.: Moscow, Russia.

42. Bernstein, S.; Campbell, E.C. Nuclear anomalous dispersion in ^{57}Fe by the method of total reflection. *Phys. Rev.* **1963**, 1625–1633. [CrossRef]

43. Andreeva, M.A.; Rosete, C. Theory of reflection from Mössbauer mirror. Taking account of laminar variation in the parameters of the hyperfine interactions close to the surface. *Vestnik Moskovskogo Universiteta Fizika* **1986**, *41*, 57–62.

44. Röhlsberger, R. Nuclear resonant scattering of synchrotron radiation from thin films. *Hyperfine Interact.* **1999**, *123/124*, 455–479. [CrossRef]

45. Andreeva, M.; Gupta, A.; Sharma, G.; Kamali, S.; Okada, K.; Yoda, Y. Field induced spin reorientation in [Fe/Cr]$_n$ multilayers studied by nuclear resonance reflectivity. *Phys. Rev. B* **2015**, *92*, 134403. [CrossRef]

46. Andreeva, M.A.; Lindgren, B.; Panchuck, V. REFTIM. Available online: http://www.esrf.eu/Instrumentation/software/data-analysis/OurSoftware/REFTIM-1 (accessed on 25 August 2009).

47. Baron, A.Q.R.; Arthur, J.; Ruby, S.L.; Chumakov, A.I.; Smirnov, G.V.; Brown, G.S. Angular dependence of specular resonant nuclear scattering of x rays. *Phys. Rev. B* **1994**, *50*, 10354–10357. [CrossRef]

48. Andreeva, M.A.; Lindgren, B. Standing waves and reflectivity from an ultrathin layer. *JETP Lett.* **2002**, *76*, 704–706. [CrossRef]

49. Andreeva, M.; Lindgren, B. Nuclear resonant spectroscopy at Bragg reflections from periodic multilayers: Basic effects and applications. *Phys. Rev. B* **2005**, *72*, 125422. [CrossRef]

50. Andreeva, M.A.; Baulin, R.A.; Repchenko, Y.L. Standing wave approach in the theory of x-ray magnetic reflectivity. *arXiv*, 2018; arXiv:1804.05104.

51. Andreeva, M.A.; Baulin, R.A.; Borisov, M.M.; Gan'shina, E.A.; Kurlyandskaya, G.V.; Mukhamedzhnov, E.K.; Repchenko, Y.L.; Svalov, A.V. Magnetic Dichroism in the Reflectivity of Linearly Polarized Synchrotron Radiation from a Ti(10 nm)/Gd$_{0.23}$Co$_{0.77}$(250 nm)/Ti(10 nm) Sample. *J. Exp. Theor. Phys.* **2018**, *125*, 802–810. [CrossRef]

52. Henke, B.L. Ultrasoft-X-Ray Reflection, Refraction, and Production of Photoelectrons (100-1000-eV Region). *Phys. Rev. A* **1972**, *6*, 94–104. [CrossRef]

53. Bedzyk, M.J.; Bommarito, G.M.; Schildkraut, J.S. X-ray standing waves at a reflecting mirror surface. *Phys. Rev. Lett.* **1989**, *62*, 1376–1379. [CrossRef] [PubMed]

54. Chumakov, A.I.; Smirnov, G.V. Mössbauer spectroscopy of conversion electrons: Determining the range of depths that can be analyzed by nondestructive depth profiling. *Sov. Phys. JETP* **1985**, *62*, 1044.

55. Andreeva, M.A.; Baulin, R.A.; Chumakov, A.I.; Rüffer, R.; Smirnov, G.V.; Babanov, Y.A.; Devyaterikov, D.I.; Goloborodsky, B.Y.; Ponomarev, D.A.; Romashev, L.N.; et al. Field-temperature evolution of the magnetic state of [Fe(1.2 Å)/Cr(10.5 Å)]*30 sample by Mössbauer reflectometry with synchrotron radiation. *J. Magn. Magn. Mater.* **2017**, *440*, 225–229. [CrossRef]

56. Rüffer, R.; Chumakov, A.I. Nuclear Resonance Beamline at ESRF. *Hyperfine Interact.* **1996**, *97/98*, 589–604. [CrossRef]

57. Chumakov, A.I.; Sergeev, I.; Celse, J.-P.; Rüffer, R.; Lesourd, M.; Zhang, L.; Sanchez del Rio, M. Performance of a silicon monochromator under high heat load. *J. Synchrotron Radiat.* **2014**, *21*, 315–324. [CrossRef] [PubMed]

58. Ragulskaya, A.V.; Andreeva, M.A.; Rogachev, M.A.; Yakunin, S.N. The investigation of [Fe/Cr] multilayer by GISAXS. *Superlattices Microstruct.* **2019**, *125*, 16–25. [CrossRef]

59. Andreeva, M.A.; Chumakov, A.I.; Smirnov, G.V.; Babanov, Y.A.; Devyaterikov, D.I.; Goloborodsky, B.Y.; Ponomarev, D.A.; Romashev, L.N.; Ustinov, V.V.; Rüffer, R. Striking anomalies in shape of the Mössbauer spectra measured near "magnetic" Bragg reflection from [Fe/Cr] multilayer. *Hyperfine Interact.* **2016**, *237*, 1–9. [CrossRef]

60. Müller, L.; Hartmann, G.; Schleitzer, S.; Berntsen, M.H.; Walther, M.; Rysov, R.; Roseker, W.; Scholz, F.; Seltmann, J.; Glaser, L.; et al. Note: Soft X-ray transmission polarizer based on ferromagnetic thin films. *Rev. Sci. Instrum.* **2018**, *89*, 036103. [CrossRef] [PubMed]

61. Schneider, J.R.; Nagasawa, H.; Berman, L.E.; Hastings, J.B.; Siddons, D.P.; Zulehner, W. Test of annealed Czochralski grown silicon crystals as X-ray diffraction elements with 145 keV synchrotron radiation. *Nucl. Instrum. Methods* **1989**, *A276*, 636–642. [CrossRef]

 *condensed matter*

Article

Structural Evolution of MoO$_3$ Thin Films Deposited on Copper Substrates upon Annealing: An X-ray Absorption Spectroscopy Study

Salvatore Macis [1,2,*], Javad Rezvani [2], Ivan Davoli [1], Giannantonio Cibin [3], Bruno Spataro [2], Jessica Scifo [2], Luigi Faillace [4] and Augusto Marcelli [2,5]

[1]   Department of Physics, Università di Roma Tor Vergata, via della Ricerca Scientifica 1, 00133 Rome, Italy; ivan.davoli@roma2.infn.it
[2]   Istituto Nazionale Fisica Nucleare, Laboratori Nazionali di Frascati, via Enrico Fermi 40, 00044 Frascati, Italy; Javad.rezvani@lnf.infn.it (J.R.); bruno.spataro@lnf.infn.it (B.S.); jessica.scifo@lnf.infn.it (J.S.); Augusto.Marcelli@lnf.infn.it (A.M.)
[3]   Diamond Light Source, Harwell Science and Innovation Campus, Didcot OX11 0DE, UK; giannantonio.cibin@diamond.ac.uk
[4]   Istituto Nazionale Fisica Nucleare, Sezione di Milano, Via Celoria 16, 20133 Milano, Italy; faillax81@gmail.com
[5]   Rome International Centre for Material Science Superstripes, RICMASS, via dei Sabelli 119A, 00185 Rome, Italy
*   Correspondence: salvatore.macis@roma2.infn.it

Received: 14 March 2019; Accepted: 15 April 2019; Published: 18 April 2019

Abstract: Structural changes of MoO$_3$ thin films deposited on thick copper substrates upon annealing at different temperatures were investigated via ex situ X-Ray Absorption Spectroscopy (XAS). From the analysis of the X-ray Absorption Near-Edge Structure (XANES) pre-edge and Extended X-ray Absorption Fine Structure (EXAFS), we show the dynamics of the structural order and of the valence state. As-deposited films were mainly disordered, and ordering phenomena did not occur for annealing temperatures up to 300 °C. At ~350 °C, a dominant α-MoO$_3$ crystalline phase started to emerge, and XAS spectra ruled out the formation of a molybdenum dioxide phase. A further increase of the annealing temperature to ~500 °C resulted in a complex phase transformation with a concurrent reduction of Mo^{6+} ions to Mo^{4+}. These original results suggest the possibility of using MoO$_3$ as a hard, protective, transparent, and conductive material in different technologies, such as accelerating copper-based devices, to reduce damage at high gradients.

Keywords: molybdenum; TM oxides; XAFS; thin films

---

1. Introduction

Molybdenum-based oxides are amongst the most adaptable and functional oxides due to their unique characteristics and tunable properties [1–3]. Molybdenum trioxide (MoO$_3$) is one of the thermodynamically stable molybdenum oxides, with the orthorhombic crystal structure α-MoO$_3$ [3,4]. The latter phase consists of a set of layers, each one containing distorted MoO$_6$ octahedra, characterized by three different oxygen sites: a single, a double, and a triple shared site. The dipole nature of the α-MoO$_3$ layers is at the origin of its relatively high work function (WF) of about 6.5 eV [1,5]. Moreover, despite its insulator nature and high WF, previous studies pointed out that thin MoO$_3$ films may exhibit a conductive behavior in the presence of defects and oxygen vacancies [6–10] or via interaction with a metallic substrate such as copper [7].

MoO$_3$ is used in solar cells, batteries, and organic light-emitting diodes (OLEDs), and it is a promising material for protective coatings of accelerating radiofrequency (RF) cavities [5]. In fact, due to the low WF, the performance and the lifetime of a copper RF cavity are strongly affected by breakdown phenomena and thermal stress generated by electron emission from the surface [11]. A high WF conductive coating could be used to reduce these detrimental phenomena, extending the lifetime of the device and allowing it to operate at higher electric fields [1,5,11,12]. Hence, MoO_3 coating can significantly enhance the electronic and mechanical properties of copper-based devices via its high WF and relatively higher hardness, compared with copper [5,13–15], without affecting the surface conductivity. Because the thermal evaporation deposition produces disordered and poorly adhesive MoO_3 coatings [5], with the aim to obtain ordered MoO_3 coatings, we tried to optimize the annealing procedure. This method minimized the formation of MoO_2, which must be avoided, since the slightly off-axis position of Mo atoms in the MoO_2 phase causes the lowering of the WF (~4.6 eV) [3,16]. To establish the optimal annealing temperature for an ordered MoO_3 film on a copper substrate, we investigated the structural evolution of annealed MoO_3 films by X-ray Absorption Spectroscopy (XAF) in X-ray Absorption Near-Edge Structure (XANES) and in the Extended X-ray Absorption Fine Structure (EXAFS) regions [17–20].

2. Materials and Methods

Molybdenum trioxide films were deposited in a dedicated vacuum sublimation set up on 5 mm-thick copper substrates [16]. Molybdenum trioxide powder (99.97% trace metal basis, Sigma-Aldrich®, St. Louis, MO, USA) was heated up to 600 °C in a tungsten crucible inside the evaporation chamber with a base pressure of 10^{-5} mbar. In order to anneal the coating without oxidizing the copper substrate, a heat treatment in a low vacuum environment was performed, with a base pressure of 5×10^{-1} mbar. We also considered a fast heating procedure to minimize the Mo reduction process and to further reduce copper oxidization. This setup allowed us to reach a temperature of up to 500 °C in less than 10 min, with a constant heating rate of ~1 °C/s. After a brief temperature decrease, the samples were exposed to air at 200 °C in order to increase the amount of oxygen in the film [16].

X-Ray Absorption Spectroscopy (XAS) measurements were performed at the B08 beamline at the European Synchrotron Radiation Facility (ESRF) in Grenoble, which works at the energy of 6 GeV and with a current of ~200 mA in the top-up mode. The B08 beamline is the Italian CRG (LISA), optimized for X-ray absorption measurements. Its optical layout covers a wide range of energies, from 5 to 40 keV, and, with the Si(111) crystals, delivers a flux to the sample of ~10^{11} ph/s within a spot of ~200 µm [21]. The acquisition at the Mo K-edge (19,999 eV) spectra was performed in the energy step-scan mode, and the fluorescence signal was collected by a 12-element Ge detector. The measurements were carried out with an energy resolution in the order of 2 eV, and the scan steps were set to 0.5 eV and 1 eV in the XANES and the EXAFS region, respectively.

X-ray Absorption Spectroscopy (XAS) analysis were performed using the software package FEFF6 (Seattle, WA, USA) [22]. Background subtraction and normalization of the absorption spectra were performed by fitting the pre-edge region with a first-order polynomial, and the spectrum after the edge with a cubic polynomial. The Mo K edge absorption threshold was associated with the first maximum of the first derivative.

3. Results and Discussion

The comparison of XANES spectra at the Mo K edge of the 250 nm MoO_3 films deposited on a copper substrate and annealed at 300 °C, 350 °C, and 500 °C is shown in Figure 1, along with the MoO_3 reference spectrum. The latter has a well-defined pre-edge peak at 20,005.6 eV (peak A), with two other major components at 20,025.7 eV and 20,037.2 eV (peaks B and C). The broadened features of the as-deposited film, in comparison with the reference spectra of the α-MoO_3 spectrum, confirmed that this film was mainly disordered [23–25]. Within the initial annealing stage at 300 °C, the ordering process (flagged by peak C) started to appear, although the film remained disordered. Increasing the annealing temperature to 350 °C triggered an ordering process within the film, leading to the observation of the typical MoO_3 features (peaks B and C). At the annealing temperature of 500 °C, the spectrum indicated the occurrence of a phase transformation: a clear shift of 3.8 eV of the third component at the energy of 20,033.4 eV (peak C) was observed. This dynamic can be also associated with the partial reduction of Mo ions due to oxygen loss [23,26,27]. A closer look to the XANES spectra also revealed changes in the pre-edge peak as a function of the annealing temperature. Above 350 °C, we observed a broadening and a lowering of the pre-edge component that can be associated with the partial reduction of Mo^{6+} ions to Mo^{4+}. Indeed, the reduction of the pre-edge intensity, that is a probe of the local and partial empty density of states around Mo atoms, was due to the direct 1s to 4d quadrupole-allowed transitions and to the dipole-allowed 1s hybridized (5p,4d) states.

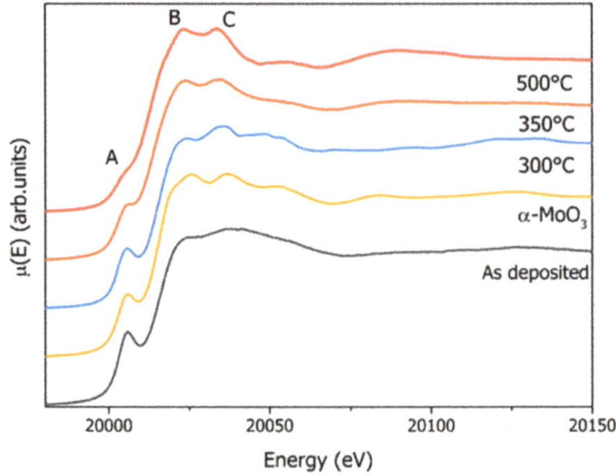

Figure 1. Comparison of X-ray Absorption Near-Edge Structure (XANES) spectra at the molybdenum K edge for the standard α-MoO_3 powder (yellow), the as-deposited MoO_3 film 250 nm thick (grey), and those annealed at 300 °C (blue), 350 °C (orange), and 500 °C (red).

In order to probe the dependence of the structural changes of these oxide films on the annealing temperature, a more detailed XAS analysis was carried out. Figure 2 shows the comparison of the pseudo-radial distribution of the distances in these samples, obtained from the Fourier transform (FT) of the EXAFS X(k) from the as-deposited film, the annealed samples, and the α-MoO_3 reference.

The as-deposited film and the one annealed at 300 °C exhibited only a large peak at around 1.7 Å, corresponding to the nearest oxygen neighbors (Mo–O) without additional shells. This is in agreement with previous XANES spectra of these same samples that pointed out the presence of disordered phases [23]. On the other hand, a structural ordering of the coating was observed at higher annealing temperatures (> 350 °C). As shown in Figure 2, at 350 °C, a split of the first peak appeared, similar to the spectrum of the reference, and a second major component associated with the nearest Mo neighbors (Mo–Mo second shell distance) also emerged at around 3.5 Å. A comparison with the α-MoO_3 spectrum confirmed the formation of the dominant MoO_3 phase, in agreement with the XANES spectra. At 500 °C, the FT was different from that of the film annealed at 350 °C, showing a single strong peak due to changes in the first Mo–O shells and in good agreement with the FT of the MoO_2 phase [27]. The Mo–Mo peak at the distance of ~2.5 Å also appeared, in agreement with the edge-sharing octahedra characteristic of the distorted rutile structure of the MoO_2 phase [27].

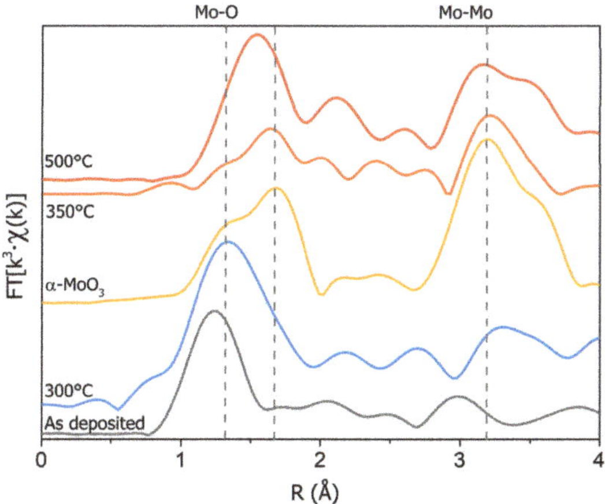

Figure 2. Fourier transform (FT) of the $k^3 \cdot X(k)$ signal among the disordered MoO_3 film (gray), the α-MoO_3 (yellow), the 300 °C (blue), 350 °C (orange), and 500 °C (red) annealed MoO_3 films deposited on Cu and the α-MoO_3 powder (yellow). The vertical continuous lines refer to the position of MoO_3 main peaks. It is clear that at 500 °C, the distribution is mainly related to MoO_2, while at 350 °C, it corresponds to the α-MoO_3 phase.

The EXAFS data of the α-MoO_3 reference and annealed films were fitted using the FEFF package with MoO_3 and MoO_2 reference crystalline structures, respectively (see Figure 3). Structural refinements were performed by minimizing the difference of the raw absorption spectra with the simulation, including the structural oscillations, $\chi(\kappa)$, and a suitable background function. The fit was performed in two steps: first, the E_0 and S_0^2 parameters fitting only the first shell were calculated. In the second step, only the distances (R) and σ^2 were left free. During this procedure, we used constant coordination numbers obtained from MoO_3 and MoO_2 crystal structures. The results (see Table 1) of the 350 °C annealed sample returned distances in good agreement with the α-MoO_3 structure, with a reasonable mean-square relative displacement ($\sigma^2 = 0.0015$ Å^2).

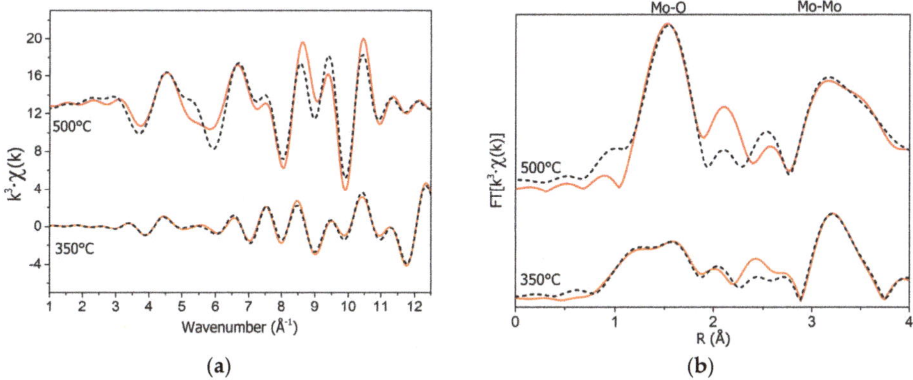

Figure 3. (**Left**) Comparison of the $k^3 \cdot X(k)$ and the FT (**right**) of annealed MoO_3 films deposited on Cu at 350 °C (orange) and 500 °C (red). Corresponding fits are the black dashed lines. The 350 °C signal was fitted with the α-MoO_3 structure, while the 500 °C signal was fitted with the orthorhombic MoO_2 structure.

Table 1. Best-fit values of the samples from Extended X-ray Absorption Fine Structure (EXAFS) spectra. For the 350 °C sample, the fit was obtained from a refinement of the sum of theoretical EXAFS calculated for the orthorhombic MoO_3 structure (k range = 3–12.5 Å^{-1}, R = 0.5–4.2 Å, 8 single scattering paths, 16 free parameters, and two fixed, r-factor = 0.027, S_0^2 = 0.6) and a monoclinic MoO_2 structure for the 500 °C sample (k range = 3–12.5 Å^{-1}, R = 0.5–4.2 Å, 6 single scattering paths, 12 free parameters, and two fixed, r-factor = 0.021, S_0^2 = 0.8). The coordination numbers are those of the MoO_3 and MoO_2 structures.

α-MoO_3 Reference				350 °C Annealing				500 °C Annealing			
Shell	**CN**	**R(Å)**	**$\sigma^2(\text{Å}^2)$**	**Shell**	**CN**	**R(Å)**	**$\sigma^2(\text{Å}^2)$**	**Shell**	**CN**	**R(Å)**	**$\sigma^2(\text{Å}^2)$**
Mo-O	1	1.70 ± 0.05	0.0024 ± 0.0004	**Mo-O**	1	1.70 ± 0.08	0.0015 ± 0.0012	**Mo-O**	2	1.99 ± 0.07	0.0037 ± 0.0011
Mo-O	1	1.78 ± 0.10	0.0023 ± 0.0003	**Mo-O**	1	1.80 ± 0.08	0.0013 ± 0.0004	**Mo-O**	4	2.02 ± 0.05	0.0031 ± 0.0016
Mo-O	2	1.99 ± 0.06	0.0025 ± 0.0002	**Mo-O**	2	2.07 ± 0.10	0.0021 ± 0.0010	**Mo-Mo**	2	3.17 ± 0.06	0.0041 ± 0.0012
Mo-O	1	2.22 ± 0.04	0.0022 ± 0.0004	**Mo-O**	1	2.22 ± 0.11	0.0022 ± 0.0017	**Mo-O**	4	3.42 ± 0.05	0.0029 ± 0.0010
Mo-O	1	2.31 ± 0.03	0.0022 ± 0.0003	**Mo-O**	1	2.30 ± 0.05	0.0016 ± 0.0009	**Mo-Mo**	8	3.73 ± 0.07	0.0037 ± 0.0007
Mo-Mo	2	3.41 ± 0.03	0.0027 ± 0.0003	**Mo-Mo**	2	3.38 ± 0.05	0.0027 ± 0.0002	**Mo-O**	4	4.05 ± 0.08	0.0035 ± 0.0005
Mo-Mo	2	3.75 ± 0.02	0.0025 ± 0.0002	**Mo-Mo**	2	3.73 ± 0.10	0.0019 ± 0.0002				
Mo-Mo	2	4.02 ± 0.09	0.0025 ± 0.0010	**Mo-Mo**	2	4.08 ± 0.04	0.0019 ± 0.0005				

On the contrary, the fit of the sample annealed at 500 °C showed a dominant MoO_2 phase, with only the distances of the MoO_2 structure ruling out the contribution of MoO_3 phases. The small deviations of the fit from the experimental spectrum can be associated with the presence of minor amounts of other non-stoichiometric oxide phases.

4. Conclusions

The structural evolution of MoO_3 films deposited on copper after annealing was investigated with XAS spectroscopy. At lower annealing temperatures, the experimental spectra did not change, showing that the as-deposited films were disordered up to ~300 °C. For the annealing procedure in a low vacuum regime (0.5 mbar) at around 350 °C, the spectra shows that the film underwent an ordering process, with the formation of the α-MoO_3 phase, as confirmed by the EXAFS analysis. Moreover, XANES spectra at 500 °C showed that the molybdenum ions underwent a decrease of the pre-edge intensity, in agreement with a reduction from aMo^{6+} to a Mo^{4+} valence state. The fit of the EXAFS spectra showed that the phase of these films could be mainly associated with the presence of the MoO_2 phase. These results represent a first important advancement for many foreseen applications of these coatings, in particular for compact RF devices made of copper.

Author Contributions: Conceptualization, methodology, S.M., A.M., and I.D.; data analysis, S.M., J.R., and G.C.; Manuscript preparation: S.M., A.M., and J.R.; S.M., A.M., J.R., I.D., G.C., B.S., J.S., L.F. have reviewed the manuscript.

Funding: We acknowledge the INFN for the support within DEMETRA and NUCLEAAR, two projects funded by the INFN Vth Committee. We also acknowledge the financial support of the Bilateral Cooperation Agreement between Italy and Japan of the *Italian Ministry of Foreign Affairs and of the International Cooperation* (MAECI) in the framework of the project of major relevance N. PGR0072.

Acknowledgments: We acknowledge ESRF and CNR-IOM for providing beamtime (proposal n. 08-01 1056) at LISA, the Italian beamline at ESRF. We acknowledge A. Bianconi for many fruitful discussions and P. De Padova and M. Lucci for their contribution to the development of the annealing procedure and to the experimental setup.

Conflicts of Interest: The authors declare no conflict of interest.

References

1. Guo, Y.; Robertson, J. Origin of the high work function and high conductivity of MoO_3. *Appl. Phys. Lett.* **2014**, *105*, 222110. [CrossRef]
2. Meyer, J.; Hamwi, S.; Kröger, M.; Kowalsky, W.; Riedl, T.; Kahn, A. Transition metal oxides for organic electronics: energetics, device physics and applications. *Adv. Mater.* **2012**, *24*, 5408–5427. [CrossRef]

3. Scanlon, D.O.; Watson, G.W.; Payne, D.J.; Atkinson, G.R.; Egdell, R.G.; Law, D.S.L. Theoretical and experimental study of the electronic structures of MoO_3 and MoO_2. *J. Phys. Chem. C* **2010**, *114*, 4636–4645. [CrossRef]

4. Hanson, E.D.; Lajaunie, L.; Hao, S.; Myers, B.D.; Shi, F.; Murthy, A.A.; Dravid, V.P. Systematic study of oxygen vacancy tunable transport properties of few-layer MoO_{3-x} enabled by vapor-based synthesis. *Adv. Funct. Mater.* **2017**, *27*, 1605380. [CrossRef]

5. Macis, S.; Aramo, C.; Bonavolontà, C.; Cibin, G.; D'Elia, A.I.; Davoli, M.; De Lucia, M.; Lucci, S.; Lupi, M.; Miliucci, A.; et al. MoO_3 films Grown on polycrystalline Cu: Morphological, structural, and electronic properties. *J. Vac. Sci. Technol. A* **2019**, *37*, 021513. [CrossRef]

6. De Castro, I.A.; Datta, R.S.; Ou, J.Z.; Castellanos-Gomez, A.S.; Sriram, T.D.; Kalantar-zadeh, K. Molybdenum oxides–from fundamentals to functionality. *Adv. Mat.* **2017**, *29*, 1701619. [CrossRef]

7. Greiner, M.T.; Chai, L.; Helander, M.G.; Tang, W.; Lu, Z.H. Metal/metal-oxide interfaces: How metal contacts affect the work function and band structure of MoO_3. *Adv. Funct. Mater.* **2013**, *23*, 215–226. [CrossRef]

8. Lambert, D.S.; Lennon, A.; Burr, P.A. Extrinsic defects in crystalline MoO_3: Solubility and effect on the electronic structure. *J. Phys. Chem. C* **2018**, *122*, 27241–27249. [CrossRef]

9. Lambert, D.S.; Murphy, S.T.; Lennon, A.; Burr, P.A. Formation of intrinsic and silicon defects in MoO_3 under varied oxygen partial pressure and temperature conditions: An ab initio DFT investigation. *RSC Adv.* **2017**, *7*, 53810–53821. [CrossRef]

10. Akande, S.O.; Chroneos, A.; Vasilopoulou, M.; Kennou, S.; Schwingenschlögl, U. Vacancy formation in MoO_3: Hybrid density functional theory and photoemission experiments. *J. Mater. Chem. C* **2016**, *4*, 9526–9531. [CrossRef]

11. Castorina, G.; Marcelli, A.; Monforte, F.; Sarti, S.; Spataro, B. An analytical model for evaluation of the properties of metallic coatings in RF structures. *Condens. Matter* **2016**, *1*, 12. [CrossRef]

12. Marcelli, A.; Spataro, B.; Sarti, S.; Dolgashev, V.A.; Tantawi, S.; Yeremian, D.A.; Higashi, Y.; Parodi, R.; Notargiacomo, A.; Junqing, X.G.; et al. Characterization of thick conducting molybdenum films: Enhanced conductivity via thermal annealing. *Surf. Coat. Tech.* **2015**, *261*, 391–397. [CrossRef]

13. Xu, Y.; Spataro, B.; Sarti, S.; Dolgashev, V.A.; Tantawi, S.; Yeremian, A.D.; Higashi, Y.; Grimaldi, M.G.; Romano, L.; Ruffino, F. Structural and morphological characterization of Mo coatings for high gradient accelerating structures. *J. Phys. Conf. Ser.* **2013**, *430*, 012091. [CrossRef]

14. Marcelli, A.; Spataro, B.; Castorina, G.; Xu, W.; Sarti, S.; Monforte, F.; Cibin, G. Materials and breakdown phenomena: Heterogeneous molybdenum metallic films. *Condens. Matter* **2017**, *2*, 18. [CrossRef]

15. Dolgashev, V.A.; Tantawi, S.G.; Park, M.; Higashi, Y.; Spataro, B. Study of basic Rf breakdown phenomena in high gradient vacuum structures. In Proceedings of the 16th International Linear Accelerator Conference LINAC2010, Tsukuba, Japan, 12–17 September 2010; pp. 1043–1047.

16. Macis, S. Deposition and Characterization of Thin MoO_3 Films on Cu for Technological Applications. Ph.D. Thesis, Roma Tor Vergata University, Roma, Italy, 2019.

17. Bianconi, A.; Marcelli, A. Surface X-Ray Absorption near-edge structure: XANES. In *Synchrotron Radiation Research. Advances in Surface Science*; Bachrach, R.Z., Ed.; Plenum Press: New York, NY, USA, 1992; Chapter 2; Volume 1.

18. Marcelli, A. Phase separations in highly correlated materials. *Acta Phys. Polonica A* **2016**, *129*, 264–269. [CrossRef]

19. Garcia, J.; Benfatto, M.; Natoli, C.R.; Bianconi, A.; Davoli, I.; Marcelli, A. Three particle correlation function of metal ions in tetrahedral coordination determined by XANES. *Solid State Commun.* **1986**, *58*, 595–599. [CrossRef]

20. Giuli, G.; Paris, E.; Wu, Z.; Brigatti, M.F.; Cibin, G.; Mottana, A.; Marcelli, A. Experimental and theoretical XANES and EXAFS study of tetra-ferriphlogopit. *Eur. J. Miner.* **2001**, *13*, 1099–1108. [CrossRef]

21. d'Acapito, F.; Lepore, G.O.; Puri, A.; Laloni, A.; la Manna, F.; Dettona, E.; de Luisa, A.; Martin, A. The LISA beamline at ESRF. *J. Synchrotron Radiat.* **2019**, *26*, 551–558.

22. Zabinsky, S.I.; Rehr, J.J.; Ankudinov, A.; Albers, R.C.; Eller, M.J. Multiple scattering calculations of X-ray absorption spectra. *Phys. Rev. B* **1995**, *52*, 2995. [CrossRef]

23. Kopachevska, N.S.; Melnyk, A.K.; Bacherikova, I.V.; Zazhigalov, V.A.; Wieczorek-Ciurowa, K. Determination of molybdenum oxidation state on the mechanochemically treated MoO_3. *Хіміяфізиката Технологія Поверхні* **2015**, *6*, 474–480, ISSN 2079-1704.

24. Di Cicco, A.; Bianconi, A.; Coluzza, C.; Rudolf, P.; Lagarde, P.; Flank, A.M.; Marcelli, A. XANES study of structural disorder in amorphous silicon. *J. Non-Cryst. Solids* **1990**, *116*, 27–32. [CrossRef]
25. Bianconi, A.; Garcia, J.; Marcelli, A.; Benfatto, M.; Natoli, C.R.; Davoli, I. Probing higher order correlation functions in liquids by XANES (X-ray absorption Near Edge Structure). *J. Phys. Colloq.* **1985**, *46*, 101–106. [CrossRef]
26. Ressler, T.; Jentoft, R.E.; Wienold, J.; Gu1nter, M.M.; Timpe, O. In situ XAS and XRD studies on the formation of Mo suboxides during reduction of MoO_3. *J. Phys. Chem. B* **2000**, *104*, 6360–6370. [CrossRef]
27. Ressler, T.; Wienold, J.; Jentoft, R.E. Formation of bronzes during temperature-programmed reduction of MoO_3 with hydrogen—An in situ XRD and XAFS study. *Solid State Ion.* **2001**, *141*, 243–251. [CrossRef]

Article

X-Ray Absorption Spectroscopy Measurements of Cu-ProIAPP Complexes at Physiological Concentrations

Emiliano De Santis [1,2], Emma Shardlow [3], Francesco Stellato [2,*], Olivier Proux [4], Giancarlo Rossi [1,2,5], Christopher Exley [3] and Silvia Morante [1,2]

[1]   Dipartimento di Fisica, Università di Roma "Tor Vergata", Via della Ricerca Scientifica, I-00133 Roma, Italy; edesantis@roma2.infn.it (E.D.S.); rossig@roma2.infn.it (G.R.); morante@roma2.infn.it (S.M.)

[2]   Istituto Nazionale di Fisica Nucleare (INFN), Sezione di Roma 2, Via della Ricerca Scientifica, I-00133 Roma, Italy

[3]   The Birchall Centre, Lennard-Jones Laboratories, Keele University, Staffordshire ST5 5BG, UK; e.shardlow@keele.ac.uk (E.S.); c.exley@keele.ac.uk (C.E.)

[4]   Observatoire des Sciences de l'Univers de Grenoble, UMS 832 CNRS-Université Grenoble Alpes, 38041 Grenoble, France; olivier.proux@esrf.fr

[5]   Centro Fermi-Museo Storico della Fisica e Centro Studi e Ricerche "Enrico Fermi", 00184 Roma, Italy

*   Correspondence: francesco.stellato@roma2.infn.it

Received: 12 December 2018; Accepted: 15 January 2019; Published: 18 January 2019

Abstract: The amyloidogenic islet amyloid polypeptide (IAPP) and the associated pro-peptide ProIAPP$_{1-48}$ are involved in cell death in type 2 diabetes mellitus. It has been observed that interactions of this peptide with metal ions have an impact on the cytotoxicity of the peptides as well as on their deposition in the form of amyloid fibrils. In particular, Cu(II) seems to inhibit amyloid fibril formation, thus suggesting that Cu homeostasis imbalance may be involved in the pathogenesis of type 2 diabetes mellitus. We performed X-ray Absorption Spectroscopy (XAS) measurements of Cu(II)-ProIAPP complexes under near-physiological (10 μM), equimolar concentrations of Cu(II) and peptide. Such low concentrations were made accessible to XAS measurements owing to the use of the High Energy Resolved Fluorescence Detection XAS facility recently installed at the ESRF beamline BM16 (FAME-UHD). Our preliminary data show that XAS measurements at micromolar concentrations are feasible and confirm that ProIAPP$_{1-48}$-Cu(II) binding at near-physiological conditions can be detected.

Keywords: X-ray absorption spectroscopy; amylin; high energy resolution fluorescence detection

---

1. Introduction

Type 2 diabetes mellitus (T2DM) is one of the most common chronic diseases, affecting over 300 million people worldwide. Amylin or Islet Amyloid PolyPeptide (IAPP), is a peptide composed of 37 amino acids that was first discovered as a constituent of amyloid deposits in the islets of Langerhans in individuals diagnosed with diabetes [1]. IAPP is highly amyloidogenic and it is this property that implicates it in the degeneration of islet β cells in diabetes [2]. The precursor to IAPP is the 67 amino acid peptide ProIAPP that upon incomplete processing leads to ProIAPP$_{1-48}$ that has also been found in amyloid deposits in diabetes [3]. Recent research implicates aberrant or incomplete processing of ProIAPP in the aetiology of diabetes [4].

While both IAPP and ProIAPP$_{1-48}$ readily form amyloids in vitro, their in vivo concentration is significantly below saturation and a burgeoning body of research is investigating this conundrum [5]. The aggregation of super-saturated concentrations of IAPP is influenced by aluminium [6,7], iron [7], zinc [7,8] and copper [7]. It remains equivocal as to whether Al(III), Fe(III) and Zn(II) promote amyloid

(β sheet) formation while it is clear that Cu(II) prevents IAPP from assembling into β-sheet structures [7] as recently confirmed [9–13]. ProIAPP$_{1-48}$ forms amyloid less readily than IAPP and while there are few data on its interactions with metals it is also the case that Cu(II) prevents ProIAPP$_{1-48}$ from forming β-sheets structures more prone to amyloidogenesis [12,14,15]. A sketch of the process leading to the formation of ProIAPP$_{1-48}$ and IAPP and of their effect on islet β cells is depicted in Figure 1.

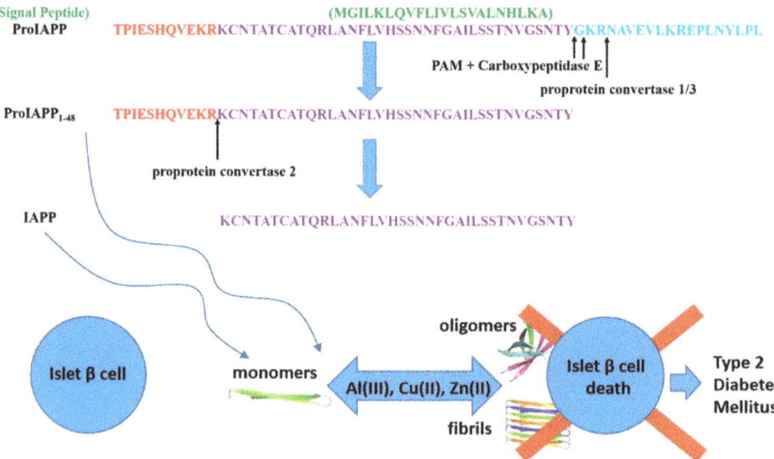

Figure 1. In the top part of the figure the amino acid sequences of ProIAPP, ProIAPP$_{1-48}$ and IAPP are given. In the bottom part we sketch the path leading to β cells death (hence to T2DM) as a consequence of fibrillation processes of ProIAPP$_{1-48}$ and IAPP monomers promoted by metal ions.

It is widely believed that the cytotoxicity of IAPP and ProIAPP$_{1-48}$ is related to their propensity to form toxic oligomers during the early stages of amyloid formation [4] and it has been suggested that metals and specifically Cu(II) potentiate toxicity through stabilisation of these oligomeric forms [12,16].

It appears therefore to be of the utmost biological and, in perspective, medical importance, to unravel the detailed interaction mechanism between metal ions and amyloidogenic islet peptides. X-ray Absorption Spectroscopy (XAS) is the technique of election to selectively obtain, at atomic resolution, information about the metals environment even in the non-crystalline systems of biological interest. Indeed, XAS has been proven to be particularly useful for the study of the metal binding mode in a number of cases related to amyloidogenic proteins [17–20].

As recalled above, both IAPP and ProIAPP$_{1-48}$ form amyloids in vitro but their typical concentration in vivo is significantly lower than the one used in in vitro experiments [21]. Since the behaviour of these peptides in the presence of metal ions does not only depend on the metal/peptide concentration ratio but can also be influenced by their absolute concentration, it is important to perform experiments as close as possible to the physiological conditions. This means that we need to perform experiments in systems where the metal concentration is in the micromolar range.

To tackle the severe signal-to-noise ratio problem associated with such low concentrations, we took advantage of the possibilities offered by the recently installed High Energy Resolved Fluorescence Detection (HERFD) XAS line, operational at the FAME-UHD beamline at ESRF (Grenoble, France) [22]. The interesting features of this high resolution XAS measurement scheme have already been successfully used to study biological samples, such as Fe-containing systems [23], Mo K-edge XAS in nitrogenase [24], the binuclear Fe centre in hydrogenase [25] and Cu organic compounds [26]. Other methods such as X-ray Emission Spectroscopy can be implemented easily using this kind of high resolution measurement, with great interest in the studies of metalloproteins [27].

With this work we pushed the limits of the technique to the study of ultra-highly diluted metal-protein complexes. The qualitative, biologically significant results we show on the metal-ProIAPP complexes are therefore intended to exhibit the potential of the HERFD-XAS technique in this context.

2. Materials and Methods

ProIAPP$_{1-48}$ fragments were synthesised using an Applied Biosystems 433A peptide synthesiser through the application of standard FMOC-based solid phase methodology. The peptide amino acid sequence is TPIESHQVEKRKCNTATCATQRLANFLVHSSNNFGAILSSTNVGSNTY. It is worth recalling that the ProIAPP$_{1-48}$ fragment is an intermediate form in the process that leads to the formation of the 37 residues long IAPP from the 67 amino acid long ProIAPP. A schematic view of the proteolytic cleavages that lead to the formation of ProIAPP$_{1-48}$ and IAPP is given in Figure 1.

Purification of the peptide was performed with the help of RP HPLC on a POROS 20R2 column using water/acetonitrile mixtures buffered with 0.1% TFA. The peptide content of the purified material (77%) was determined by quantitative amino acid analysis and lyophilised aliquots were stored at -80 °C prior to the preparation of peptide stock solutions. Peptide stocks were prepared to a final concentration of *ca* 150 μM via the addition of ultrapure water (<0.067 μS/cm) to thawed peptide lyophilisates. This stock was then used to prepare smaller individual volumes of the peptide in order to achieve the final concentrations of peptide included in the following experiments and these aliquots were stored at -20 °C until required. Thawed peptide aliquots were introduced into modified Krebs-Henseleit (KH) buffers (pH 7.4 $\pm$ 0.05) [12] with or without the respective metals (added from certified stocks (Perkin-Elmer) at the required total metal concentrations to give final peptide concentration of 10 μM.

Sub-stoichiometric, 9 μM Cu(II) ions are added thus minimizing the amount of free Cu(II) in solution. In order to investigate the effect on the Cu(II) binding mode of the presence of other metals, like Al(III) and Zn(II) ions at different concentrations, we added the second metal at a concentration of either 50 or 1500 μM. Including Cu(II) in buffer as a blank. In all we have prepared and subjected to XAS measurements the six samples listed in Table 1.

Table 1. List of the measured samples. Sample name is given in column 1, peptide concentration in column 2, Cu, Zn and Al concentrations in columns 3, 4 and 5, respectively.

Sample(μm)	[Peptide] (μm)	[Cu] (μm)	[Zn] (μm)	[Al] (μm)
Cu buffer	0	1600	0	0
Cu-ProIAPP$_{1-48}$	10	9	0	0
(Cu + Zn^{Low})-ProIAPP$_{1-48}$	10	9	50	0
(Cu + Al^{Low})-ProIAPP$_{1-48}$	10	9	0	50
(Cu + Zn^{High})-ProIAPP$_{1-48}$	10	9	1500	0
(Cu + Al^{High})-ProIAPP$_{1-48}$	10	9	0	1500

Copper K-edge (8.979 keV) XANES measurements were performed on the BM16 beamline (CRG FAME-UHD) at the ESRF (Grenoble, France) [22]. The main optical elements of the beamline were a two-crystal Si(220) monochromator located between two Rh-coated mirrors. The beam size on the sample was around 200×100 μm^2 (H $\times$ V, FWHM) thanks to the sagittal focus of the 2nd crystal of the monochromator and the vertical one of the 2nd mirror. The 1st crystal of the monochromator is liquid nitrogen cooled in order to limit its thermal bump due to the incoming photons thus increasing the energy resolution of the monochromatic beam. Energy calibration was done by setting the 1st maximum of the 1st derivative of the copper metallic foil absorption spectrum to 8.979 keV. A 5-crystal analyser spectrometer on a Johann-type geometry (Figure 2), equipped with Si(444) bent crystals with a 1m radius of curvature (from Crystal Analyser Laboratories, ESRF) was used for fluorescence detection. The spectrometer was aligned so as to have the crystals in Bragg conditions at the Cu K$_{\alpha 1}$ emission line photons energy. The overall instrument energy resolution (combining crystal analyser spectrometer and monochromator contributions) was established to be 0.7 eV by measuring the full width at half maximum of the elastic peak (measured by scanning the incident photon energy, with the monochromator, across the fluorescence photon energy, selected by the spectrometer). All the photons scattered by the crystals were collected using an energy-resolved silicon-drift detector, which allows discriminating the diffracted photons of interest from the other scattered ones.

For a typical biological system, in which the total radiation background (including elastic and inelastic scattering) is substantially more intense than the fluorescence line of interest, the advantage of background removal associated to the HERFD-XAS data acquisition scheme outweighs the disadvantage of having a lower total signal intensity with respect to a standard XAS measuring apparatus, such as a solid-state detector. XAS measurements of samples as diluted as those of interest here are made possible thanks to the use of this set-up that, significantly improve the signal-to-noise ratio, thus allowing nearly background-free measurements.

Data acquisition was performed using a liquid helium cryostat in order to limit sample evolution due to radiation damage under the beam.

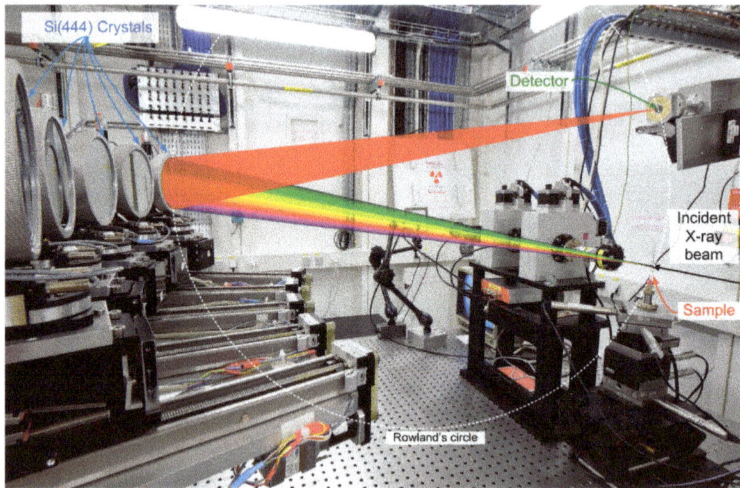

Figure 2. A picture of the FAME UHD beamline showing the HERFD acquisition geometry. In Johann geometry the sample, crystal centre and detector are on the Rowland's circle. The diameter of this circle is equal to the curvature radius of the crystals. A polyurethane balloon filled with He gas was placed between the sample, the crystals and the detector in order to minimize the air absorption of the fluorescence photons along the sample-crystal-detector path.

3. Results and Discussion

Cu K-edge XANES spectra of the sample listed in Table 1 are all gathered in Figure 3. Spectra of the Cu-ProIAPP$_{1-48}$ samples appear to be rather noisy but one has to keep in mind that Cu is present at the extremely low concentration of 9 µM (that corresponds to 0.6 ppm). To the best of our knowledge this is the first time in which XAS measurements are performed to probe the copper speciation in biological systems at bio-relevant concentration.

In the Figure 3, we compare the XANES spectrum of the Cu-ProIAPP$_{1-48}$ sample in the absence and in the presence of either Al(III) or Zn(II) ions at two different concentrations, namely 50 µM and 1500 µM, with the idea of detecting possible modifications of the Cu binding mode upon adding a second metal.

The XANES spectrum of Cu in buffer displays features typical of hydrated copper [28], while the XANES spectrum of Cu-ProIAPP$_{1-48}$ has a significantly different shape, indicating that Cu is bound to

ProIAPP$_{1-48}$. The Cu-ProIAPP$_{1-48}$ spectrum has an absorption maximum at nearly the same position as Cu in buffer, confirming that copper, as expected, is in its doubly ionized form, Cu(II).

It is interesting to note that Zn(II) and Al(III) added in solution seem to affect the Cu-ProIAPP$_{1-48}$ coordination mode in a way that depends on their concentration. In particular Zn(II) is seen to affect the XANES spectrum of Cu-ProIAPP$_{1-48}$ at high (black line), while Al(III) at low (light blue line) concentration.

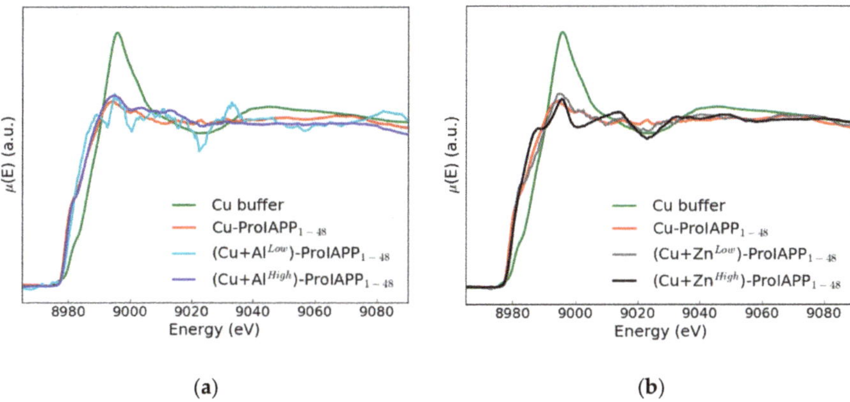

(**a**)　　　　　　　　　　　　(**b**)

Figure 3. The XANES spectra of the Cu-buffer (green line) and Cu-ProIAPP$_{1-48}$ sample in the absence of metal ions (red line) compared in panel (**a**) with the Cu-ProIAPP$_{1-48}$ spectrum in the presence of 50 μM Al(III) (light blue line) and 1500 μM Al(III) (blue line) and in panel (**b**) in the presence of 50 μM Zn(II) (grey line) and 1500 μM Zn(II) (black line).

These qualitative findings are supported by Thioflavine T (ThT) fluorescence and Dynamic Light Scattering (DLS) measurements. In Reference [14], some of the authors of the present paper observed that, at a ProIAPP$_{1-48}$ concentration of 20 μM, which is close to the one of the present study, the peptide forms ThT-fluorescence positive aggregates. In the presence of an equimolar concentration of Cu(II), the ThT-fluorescence is instead significantly reduced. In reference [21] they perform DLS measurements showing that Cu(II) has an effect on ProIAPP$_{1-48}$ aggregation in the direction of increasing the average aggregate size. The results of the experiments presented here confirm these observations and demonstrate that the effect of Cu(II) on ProIAPP$_{1-48}$ fibrillization is due to a direct binding between the Cu(II) ions and the peptide.

For what concerns the role of other metal ions, in Reference [21] the same authors show that Al(III) at equimolar concentration with the peptide does not have a detectable effect on ProIAPP$_{1-48}$ aggregate size. Although the ProIAPP$_{1-48}$ concentrations considered in Reference [21] were higher (5–60 μm) than that used in this study, this observation is in agreement with what we observe here, namely that at low concentrations Al(III) ions do not have significant effects. Moreover, in Reference [14] it was also observed that the ThT fluorescence of 20 μM ProIAPP$_{1-48}$ was unaffected by the presence of 10-fold excess of Zn(II) ion, while the fluorescence was significantly lower in the presence of an additional 10 μM Cu(II). Although these measurements are not performed in exactly the same conditions as in the present study, taken all together they contribute to form a picture in which Cu(II) most strongly affects ProIAPP$_{1-48}$ aggregation, with Al(III) and Zn(II) being able to modulate its effect.

4. Conclusions

We are well aware of the fact that the quality of the collected spectra does not allow any speculation about the structural differences of the possible different Cu(II) coordination modes induced by the presence of Al(III) or Zn(II). Nevertheless, we think that the results we obtained are relevant in two respects.

1. Though yet at a qualitative level, the existence of differences in the XANES spectral features induced by the presence of the Zn(II) or Al(III) in the Cu(II)-ProIAPP$_{1-48}$ binding mode sample is clearly established. These differences appear to be dependent (in different way) from the added ion concentration.

2. Experiments of the kind we have been able to perform at the ESRF HERFD XAS line demonstrate the general feasibility of XAS measurements on samples where the absorbing atom is present at micromolar concentration. This last fact is of special methodological relevance as it shows that it is possible to perform XAS measurements on very diluted metal-peptide complexes in physiological conditions, when raising metal ions concentration to improve the signal-to-noise ratio is not possible, as this would dramatically alter their physiological coordination mode.

In this paper we have successfully demonstrated the feasibility of XAS measurements of very diluted samples (i.e., where the absorber concentration is at the micromolar level) and proved that adding in solution either Zn(II) or Al(III) has a detectable impact on the Cu coordination. If, as a next step, one wants to arrive at a detailed description of the Cu(II)-ProIAPP$_{1-48}$ coordination mode, some kind of information about the metal site structure is required. To this end NMR, X-ray crystallography and numerical (classical and/or, *ab initio*) molecular approaches could be of much help, as demonstrated in the study of similar instances performed in References [28,29].

Concluding we would like to stress that XAS is an invaluable and irreplaceable tool when the interest is to get highly resolved (order of hundredths of Angstrom) structural information on the metal site even for very diluted samples like the ones one often finds in the case of biological systems in physiological conditions. Neither NMR (that needs isotopic labelling) nor X-ray diffraction (that requires crystals) are as informative and easily accessible as XAS in these circumstances. This point is largely acknowledged by people working in experimental groups that routinely use NMR and X-ray diffraction. Indeed, when available, NMR and/or X-ray diffraction data are often refined (confirmed) with complementary XAS measurements [30–34].

Author Contributions: Sample preparation: E.S. and C.E.; data acquisition: S.M., O.P. and F.S.; data analysis: E.D.S., S.M., O.P., G.R. and F.S.; writing: C.E., S.M., O.P., G.R., E.S. and F.S.

Funding: This work was partly supported by INAIL grant BRiC 2016 ID17/2016 and by INFN CIPE grant HPC_HTC 14064. Construction of the 5 crystals spectrometer used on BM16/FAME-UHD for this experiment was financially supported by the French National Institute for Earth Science and Astronomy of the CNRS (INSU CNRS), ANR NANOSURF (coordinator: C. Chaneac, LCMCP), ANR MESONNET (coordinator: J.Y. Bottero, CEREGE), CEREGE laboratory (Aix en Provence, France) and Labex OSUG@2020 (ANR-10-LABX-0056). The FAME-UHD project is financially supported by the French "invest for the future" EquipEx (EcoX, ANR-10-EQPX-27-01), the CEA-CNRS CRG consortium and the French National Institute for Earth Science and Astronomy of the CNRS (INSU CNRS).

Conflicts of Interest: The authors declare no conflict of interest.

References

1. Cooper, G.J.; Willis, A.C.; Clark, A.; Turner, R.C.; Sim, R.B.; Reid, K.B. Purification and characterization of a peptide from amyloid-rich pancreases of type 2 diabetic patients. *Proc. Natl. Acad. Sci. USA* **1987**, *84*, 8628–8632. [CrossRef] [PubMed]

2. Konarkowska, B.; Aitken, J.F.; Kistler, J.; Zhang, S.G.; Cooper, J. The aggregation potential of human amylin determines its cytotoxicity towards islet beta-cells. *FEBS J.* **2006**, *273*, 3614–3624. [CrossRef] [PubMed]

3. Westermark, P.; Engstrom, U.; Westermark, G.T.; Johnson, K.H.; Permerth, J.; Betsholtz, C. Islet amyloid polypeptide (IAPP) and pro-IAPP immunoreactivity in human islets of Langerhans. *Diabetes Res. Clin. Pract.* **1989**, *7*, 219–226. [CrossRef]

4. Courtade, J.A.; Klimek-Abercrombie, A.M.; Chen, Y.-C.; Patel, N.; Lu, P.Y.T.; Speake, C.; Orban, P.C.; Najafian, B.; Meneilly, G.; Greenbaum, C.J.; et al. Measurement of pro-islet amyloid polypeptide (1–48) in diabetes and islet transplants. *J. Clin. Endocrinol. Metab.* **2017**, *102*, 2595–2603. [CrossRef] [PubMed]

5. Raleigh, D.; Zhang, X.; Hastoy, B.; Clark, A. The β-cell assassin: IAPP cytotoxicity. *J. Mol. Endocrinol.* **2017**, *59*, R121–R140. [CrossRef] [PubMed]

6.   Exley, C.; Korchazhkina, O. Promotion of formation of amyloid fibrils by aluminium adenosine triphosphate (AlATP). *J. Inorg. Biochem.* **2001**, *84*, 215–224. [CrossRef]

7.   Ward, B.; Walker, K.; Exley, C. Cu(II) inhibits the formation of amylin amyloid in vitro. *J. Inorg. Biochem.* **2008**, *102*, 371–375. [CrossRef]

8.   Brender, J.R.; Hartman, K.; Nanga, R.P.R.; Popovych, N.; Bea, R.D.; Vivekanandan, S.; Marsh, E.G.N.; Ramamoorthy, A. Role of zinc in human islet amyloid polypeptide aggregation. *J. Am. Chem. Soc.* **2010**, *132*, 8973–8983. [CrossRef]

9.   Ma, L.; Li, X.; Wang, Y.; Zheng, W.; Chen, T. Cu(II) inhibits hIAPP fibrillation and promotes hIAPP-induced beta cell apoptosis through induction of ROS-mediated mitochondrial dysfunction. *J. Inorg. Biochem.* **2014**, *140*, 143–152. [CrossRef]

10. Li, H.; Ha, E.; Donaldson, R.P.; Jeremic, A.M.; Vertes, A. Rapid assessment of human amylin aggregation and its inhibition by copper (II) ions by laser ablation electrospray ionisation mass spectrometry with ion mobility separation. *Anal. Chem.* **2015**, *87*, 9829–9837. [CrossRef]

11. Riba, I.; Barran, P.E.; Cooper, G.J.S.; Unwin, R.D. On the structure of the copper-amylin complex. *Int. J. Mass Spectrom.* **2015**, *391*, 47–53. [CrossRef]

12. Mold, M.; Bunrat, C.; Goswami, P.; Roberts, A.; Roberts, C.; Taylor, N.; Taylor, H.; Wu, L.; Fraser, P.E.; Exley, C. Further insight into the role of metals in amyloid formation by $IAPP_{1-37}$ and $ProIAPP_{1-48}$. *J. Diabetes Res. Clin. Metab.* **2015**, *4*. [CrossRef]

13. Sánchez-López, C.; Cortés-Mejía, R.; Miotto, M.C.; Binolfi, A.; Fernández, C.O.; del Campo, J.M.; Quintanar, L. Copper coordination features of human islet amyloid polypeptide: The type 2 diabetes peptide. *Inorg. Chem.* **2016**, *55*, 10727–10740. [CrossRef]

14. Exley, C.; House, E.; Patel, T.; Wu, L.; Fraser, P.E. Human pro-islet amyloid polypeptide ($ProIAPP_{1-48}$) forms amyloid fibrils and amyloid spherulites in vitro. *J. Inorg. Biochem.* **2010**, *104*, 1125–1129. [CrossRef] [PubMed]

15. Exley, C.; Mold, M.; Shardlow, E.; Shuker, B.; Ikpe, B.; Wu, L.; Fraser, P.E. Copper is a potent inhibitor of the propensity for human $ProIAPP_{1-48}$ to form amyloid fibrils in vitro. *J. Diabetes Res. Clin. Metab.* **2012**, *1*, 3. [CrossRef]

16. Lee, S.J.C.; Choi, T.S.; Lee, J.W.; Lee, H.J.; Mun, D.-G.; Akashi, S.; Lee, S.-W.; Lim, M.H.; Kim, H.I. Structure and assembly mechanisms of toxic human islet amyloid polypeptide oligomers associated with copper. *Chem. Sci.* **2016**, *7*, 5398–5406. [CrossRef] [PubMed]

17. De Santis, E.; Minicozzi, V.; Proux, O.; Rossi, G.C.; Silva, K.I.; Lawless, M.J.; Stellato, F.; Saxena, S.; Morante, S. Cu(II)–Zn(II) Cross-Modulation in Amyloid–Beta Peptide Binding: An X-ray Absorption Spectroscopy Study. *J. Phys. Chem. B* **2015**, *119*, 15813–15820. [CrossRef]

18. Stellato, F.; Spevacek, A.; Proux, O.; Minicozzi, V.; Millhauser, G.; Morante, S. Zinc modulates copper coordination mode in prion protein octa-repeat subdomains. *Eur. Biophys. J.* **2011**, *40*, 1259–1270. [CrossRef]

19. Stellato, F.; Minicozzi, V.; Millhauser, G.L.; Pascucci, M.; Proux, O.; Rossi, G.C.; Morante, S. Copper–zinc cross-modulation in prion protein binding. *Eur. Biophys. J.* **2014**, *43*, 631–642. [CrossRef]

20. Stellato, F.; Fusco, Z.; Chiaraluce, R.; Consalvi, V.; Dinarelli, S.; Placidi, E.; Petrosino, M.; Rossi, G.C.R.; Minicozzi, V.; Morante, S. The effect of β-sheet breaker peptides on metal associated Amyloid-β peptide aggregation process. *Biophys. Chem.* **2017**, *229*, 110–114. [CrossRef]

21. Shardlow, E.; Rao, C.; Sattarov, R.; Wu, L.; Fraser, P.E.; Exley, C. Aggregation of the diabetes-related peptide $ProIAPP_{1-48}$ measured by dynamic light scattering. *J. Trace Elem. Med. Biol.* **2019**, *51*, 1–8. [CrossRef] [PubMed]

22. Proux, O.; Lahera, E.; Del Net, W.; Kieffer, I.; Rovezzi, M.; Testemale, D.; Irar, M.; Thomas, S.; Aguilar-Tapia, A.; Bazarkina, E.F.; et al. High-energy resolution fluorescence detected X-ray absorption spectroscopy: A powerful new structural tool in environmental biogeochemistry sciences. *J. Environ. Qual.* **2017**, *46*, 1146–1157. [CrossRef] [PubMed]

23. Castillo, R.G.; Banerjee, R.; Allpress, C.J.; Rohde, G.T.; Bill, E.; Que, L., Jr.; Lipscomb, J.D.; DeBeer, S. High-energy-resolution fluorescence-detected X-ray absorption of the Q intermediate of soluble methane monooxygenase. *JACS* **2017**, *139*, 18024–18033. [CrossRef] [PubMed]

24. Bjornsson, R.; Lima, F.A.; Spatzal, T.; Weyhermüller, T.; Glatzel, P.; Bill, E.; Einzle, O.; Neese, F.; DeBeer, S. Identification of a spin-coupled Mo(III) in the nitrogenase iron–molybdenum cofactor. *Chem. Sci.* **2014**, *5*, 3096–3103. [CrossRef]

25. Mebs, S.; Kositzki, R.; Duan, J.; Kertess, L.; Senger, M.; Wittkamp, F.; Apfel, U.P.; Happe, T.; Stripp, S.T.; Winkler, M.; et al. Hydrogen and oxygen trapping at the H-cluster of [FeFe]-hydrogenase revealed by site-selective spectroscopy and QM/MM calculations. *Biochim. Biophys. Acta Bioenerg.* **2018**, *1859*, 28–41. [CrossRef] [PubMed]

26. Vollmers, N.J.; Müller, P.; Hoffmann, A.; Herres-Pawlis, S.; Rohrmuüller, M.; Schmidt, W.G.; Gerstmann, U.; Bauer, M. Experimental and theoretical high-energy-resolution X-ray absorption spectroscopy: Implications for the investigation of the entatic state. *Inorg. Chem.* **2016**, *55*, 11694–11706. [CrossRef] [PubMed]

27. Kowalska, J.K.; Lima, F.A.; Pollock, C.J.; Rees, J.A.; DeBeer, S. A practical guide to high-resolution X-ray spectroscopic measurements and their applications in bioinorganic chemistry. *Isr. J. Chem.* **2016**, *56*, 803–815. [CrossRef]

28. La Penna, G.; Minicozzi, V.; Morante, S.; Rossi, G.C.; Stellato, F. A first-principle calculation of the XANES spectrum of Cu^{2+} in water. *J. Chem. Phys.* **2015**, *143*, 124508. [CrossRef] [PubMed]

29. Stellato, F.; Calandra, M.; D'Acapito, F.; De Santis, E.; La Penna, G.; Rossi, G.C.; Morante, S. Multi-scale theoretical approach to X-ray absorption spectra in disordered systems: An application to the study of Zn(II) in water. *Phys. Chem. Chem. Phys.* **2018**, *20*, 24775–24782. [CrossRef] [PubMed]

30. Banci, L.; Bertini, I.; Mangani, S. Integration of XAS and NMR techniques for the structure determination of metalloproteins. Examples from the study of copper transport proteins. *J. Synchrotron Radiat.* **2005**, *12*, 94–97. [CrossRef]

31. Koutmou, K.S.; Casiano-Negroni, A.; Getz, M.M.; Pazicni, S.; Andrews, A.J.; Penner-Hahn, J.E.; Al-Hashimi, H.M.; Fierke, C.A. NMR and XAS reveal an inner-sphere metal binding site in the P4 helix of the metallo-ribozyme ribonuclease P. *Proc. Natl. Acad. Sci. USA* **2010**, *107*, 2479–2484. [CrossRef] [PubMed]

32. Arcovito, A.; Moschetti, T.; D'Angelo, P.; Mancini, G.; Vallone, B.; Brunori, M.; Della Longa, S. An X-ray diffraction and X-ray absorption spectroscopy joint study of neuroglobin. *Arch. Biochem. Biophys.* **2008**, *475*, 7–13. [CrossRef] [PubMed]

33. Frankær, C.G.; Knudsen, M.V.; Norén, K.; Nazarenko, E.; Ståhl, K.; Harris, P. The structures of T6, T3R3 and R6 bovine insulin: Combining X-ray diffraction and absorption spectroscopy. *Acta Crystallogr. D* **2012**, *68*, 1259–1271. [CrossRef] [PubMed]

34. Koebke, K.J.; Ruckthong, L.; Meagher, J.L.; Mathieu, E.; Harland, J.; Deb, A.; Lehnert, N.; Policar, C.; Tard, C.; Penner-Han, J.E.; et al. Clarifying the Copper Coordination Environment in a de Novo Designed Red Copper Protein. *Inorg. Chem.* **2018**, *57*, 12291–12302. [CrossRef] [PubMed]

Article

Scintillator Pixel Detectors for Measurement of Compton Scattering

Mihael Makek *, Damir Bosnar and Luka Pavelić †

Department of Physics, Faculty of Science, University of Zagreb, Bijenička c. 32, 10000 Zagreb, Croatia; bosnar@phy.hr (D.B.); lpavelic@imi.hr (L.P.)

* Correspondence: makek@phy.hr; Tel.: +385-1-460-5572

† Current address: Institute for Medical Research and Occupational Health, Ksaverska Cesta 2, 10000 Zagreb, Croatia.

Received: 31 January 2019; Accepted: 13 February 2019; Published: 15 February 2019

Abstract: The Compton scattering of gamma rays is commonly detected using two detector layers, the first for detection of the recoil electron and the second for the scattered gamma. We have assembled detector modules consisting of scintillation pixels, which are able to detect and reconstruct the Compton scattering of gammas with only one readout layer. This substantially reduces the number of electronic channels and opens the possibility to construct cost-efficient Compton scattering detectors for various applications such as medical imaging, environment monitoring, or fundamental research. A module consists of a 4 × 4 matrix of lutetium fine silicate scintillators and is read out by a matching silicon photomultiplier array. Two modules have been tested with a ^{22}Na source in coincidence mode, and the performance in the detection of 511 keV gamma Compton scattering has been evaluated. The results show that Compton events can be clearly distinguished with a mean energy resolution of 12.2% ± 0.7% in a module and a coincidence time resolution of 0.56 ± 0.02 ns between the two modules.

Keywords: Compton camera; positron emission tomography; Compton scattering; scintillation detector; silicon photomultiplier; medical applications

---

1. Introduction

Compton scattering is a well-known process in which an incoming gamma ray is interacting with an electron, leaving a scattered gamma and a recoil electron in the final state. To detect and to reconstruct the Compton scattering fully, one needs position- and energy-sensitive detectors. The interaction point is determined from the location where the recoil electron is absorbed, while the direction of the scattered gamma is determined from its energy and the position of the absorption relative to the impact point.

The Compton scattering of gamma rays has lately received a growing interest in developing medical physics applications such as the new generation of Positron Emission Tomography (PET) devices, where several studies have shown that it has the potential to improve PET image quality, since it provides information about the gamma ray polarization, as an additional handle to improve the signal to noise ratio [1–3].

A common method of detection and reconstruction of gamma Compton scattering is to use two detector layers, the first for measurement of energy and the location of the recoil electron and the second for measuring the energy and absorption location of the scattered gamma, e.g., [4–8]. In PET, however, detectors are highly segmented and have a large coverage, as a pre-requisite to achieve a good spatial resolution and sensitivity; therefore, two highly-granular detector layers for Compton measurements would dramatically increase the cost of the apparatus. Several developments optimized for astrophysics observations use single detector layers to measure gamma ray polarization via Compton scattering [9,10], but a single-layer system PET system exploiting the gamma polarization has not been experimentally realized.

We assembled single-layer Compton detectors, and a system of two such modules was set up, as described in [11]. Each module consists of a 4 × 4 scintillator pixel array, read out on the back side by a matching array of Silicon Photomultipliers (SiPM) in a one-to-one coupling scheme. In this paper, we will show that such detectors are able to reconstruct the Compton scattering events of 511 keV gamma rays fully, as a prerequisite for gamma polarization measurements. The details of the energy, angle, and time reconstruction methods are presented in Section 2, and the detector performance is presented in Section 3.

2. Materials and Methods

2.1. Experimental Setup

A system of two detector modules labeled A and B has been set up. Each module consists of a 4 × 4 array of Lutetium Fine Silicate (LFS) scintillator pixels produced by Zecotec Inc. (Figure 1), read out by a matching array of SiPM produced by Hamamatsu (Model S13361-3050AE-04). Signals from each channel are first amplified, then digitized by fast pulse digitizers (CAEN V1743) at 1.6 GS/s and stored for offline analysis. A detailed description of the experimental setup and the detector performance is given in [11]. In the setup presented in this work, we additionally applied silicon grease as the optical coupling medium between the scintillators and the SiPMs (see Subsection 2.2 for details).

A ^{22}Na positron source (activity ≈ 1 μCi) in an aluminum case was positioned between the modules, 4 cm from the front face of each detector, providing two coincident annihilation gammas of 511 keV. The data acquisition system triggered only on events where a coincident detection occurred in both modules. The coincidence condition efficiently suppressed the random background from ^{176}Lu decays intrinsic to the scintillation material. The measurements were performed at room temperature typically in a range from 20–22 °C and at a moderate bias voltage $U_b = U_{br} + 1.6$ V, U_{br} being the SiPM breakdown voltage.

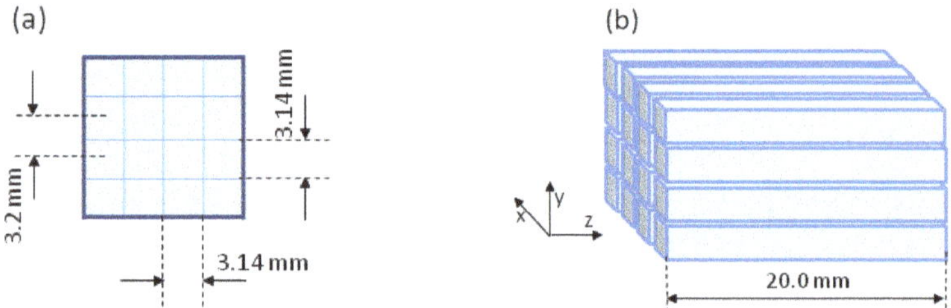

Figure 1. Schematic drawing of the scintillator pixel array: (**a**) front view; (**b**) side view.

2.2. Scintillator-SiPM Optical Coupling

In the setup presented in this work, silicon optical grease (BC-630 by Saint-Gobain) was used as the optical coupling material between the scintillator crystals and the SiPM arrays. This resulted in an increased light propagation from crystals to SiPMs, observed as an increase of signal amplitudes up to 40%, under the same operating conditions and reflected in an improvement of the energy resolution from $\Delta E/E = 12.9\%$ down to $\Delta E/E = 11.4\%$. However, this was not the case for all pixels: we found that after applying the silicon grease, some amplitudes remained at the same level as before. Consequently, small variations of energy resolution from pixel-to-pixel have been observed, resulting in the mean energy resolution of a module of 12.2% ± 0.7% (at 511 keV). Different optical coupling methods will be investigated in the future to provide a more uniform detector performance.

2.3. Signal Processing

Signals from all scintillator channels in triggered events are stored, which enables offline optimization of processing algorithms for specific applications. In our approach, the energy deposition was reconstructed by signal integration and the arrival time from the signal leading edge.

For energy reconstruction, three quantities are extracted: the average baseline, the signal amplitude, and the signal integral. The baseline is determined as the average of at least 20 samples before the signal rising edge; the amplitude is determined as the difference of the maximum sample amplitude and the baseline; and the integral is determined by summing the sampled voltages (after baseline subtraction) in the region where the signals are non-zero. An example is in Figure 2a.

To reconstruct the time of the signal arrival, we use the leading edge method: a fixed threshold is set as close to the baseline as possible to enable triggering on the very first scintillation photons, as discussed in [12]. In the presented results, the threshold is -15 mV (typical signal amplitudes reach several hundreds of mV). The algorithm searches for the sample that is just above the threshold, S, then uses linear interpolation between samples S and $S - 1$ to estimate the time when the threshold had been crossed, which is taken as the time of signal arrival (Figure 2b).

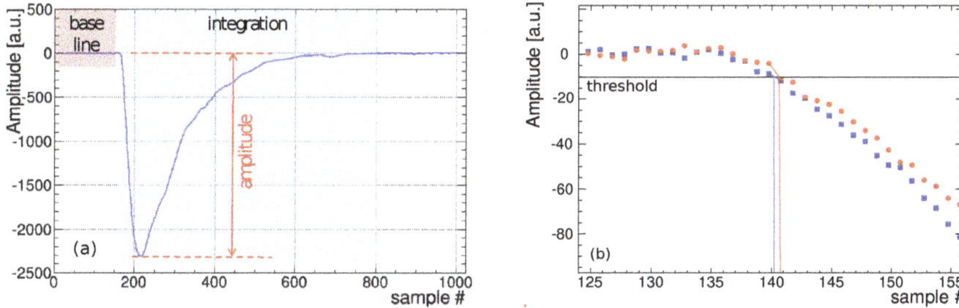

Figure 2. Examples of digitized detector signals; the distance between two samples on the X-axis corresponds to 625 ps: (**a**) the signal region where the baseline, the amplitude, and the integral are determined; (**b**) coincident signals from two detectors represented by circles (red) and squares (blue), respectively. The signal region at the beginning of the rising edge is shown. The time of the signal arrival is obtained from interpolation of the samples below and above the threshold, as marked by the red and the blue lines for the two signals, respectively.

2.4. Energy Reconstruction

2.4.1. Energy Calibration

The raw energy spectrum is obtained as the spectrum of signal integrals for each pixel. It is first corrected for non-linearity, which is present due to the finite number of micro cells per SiPM, as described in [11]. Then, it is calibrated by fitting a Gaussian to the annihilation peak and setting the corresponding value to 511 keV. For the second calibration point, we assume the zero in the integral spectrum corresponds to zero energy, inherent to the signal processing algorithm, which subtracts the baseline on an event-by-event basis. The calibration is performed for each data acquisition run, typically lasting two hours. In this time window, the temperature variations are $<0.1\ ^\circ$C and therefore have a small influence on the calibration parameters ($<0.5\%$), which is considered negligible compared to the mean energy resolution ($12.2\% \pm 0.7\%$ at 511 keV).

2.4.2. Light Sharing Correction

An energy correlation between the adjacent crystal pixels had been observed, as reported in [11]. When a significantly large signal is observed in a pixel, small signals are observed in adjacent pixels, with amplitudes proportional to the main signal amplitude, as demonstrated in Figure 3. This can be attributed to light sharing, either as light leaking through crystal sides and the Teflon reflector (0.06 mm thick) between the pixels or by light leaking at the contact with the SiPM array. When single pixel events are considered, the light sharing does not cause an energy reconstruction problem, since the energy calibration is performed using single pixel spectra. However, in two-pixel events where two adjacent pixels fire, the light sharing disguises the original energy deposition in the pixels and must be corrected for. In order to reconstruct the Compton scattering angle θ (see Section 2.5.1) correctly, it is important to determine the true energy response of the pixels. We present a method to decouple the observed energies and obtain real energy depositions in the pixels.

The energy contribution in a pixel coming form an adjacent neighbor can be described as $E_{Adj.neighbor} = \epsilon E_{main}$. Suppose two adjacent pixels are labeled 1 and 2. The measured energy responses are:

$$E_1 = E_1^o + \epsilon_{12} E_2^o \tag{1}$$

$$E_2 = E_2^o + \epsilon_{12} E_1^o \tag{2}$$

where E_1^o and E_2^o are the original energy depositions in the pixels and ϵ_{12} is the sharing fraction characteristic of that pixel pair. It follows that:

$$E_1 + E_2 = (E_1^o + E_2^o)(1 + \epsilon_{12}) \tag{3}$$

with:

$$E_1^o + E_2^o = E_\gamma \tag{4}$$

being the gamma particle energy. Hence, the energy sum of adjacent pixels 1 and 2 is boosted by a factor $(1 + \epsilon_{12})$. This is observed in Figure 4a. By combining Equation (1) to Equation (4), one obtains the original deposited energies as:

$$E_1^o = \frac{E_1 - \epsilon_{12} E_2}{1 - \epsilon_{12}^2} \tag{5}$$

$$E_2^o = \frac{E_2 - \epsilon_{12} E_1}{1 - \epsilon_{12}^2} \tag{6}$$

Since variations of ϵ_{12} are relatively small, we replace it by average module values $<\epsilon_A> = 0.060$ for detector module A and $<\epsilon_B> = 0.066$ for module B.

When two fired pixels are not adjacent neighbors, the shift of the summed energy is much smaller, as summarized in Table 1. This is also visible in Figure 4a. It can be due to indirect light sharing through an intermediate pixel or even due to dark counts from the SiPMs, which are double counted in the sum when two pixels fire. Since the effect is much smaller than with the adjacent neighbors, it is simply corrected for by scaling the measured pixel energy, E, by the same fraction:

$$E^o = (1 - \epsilon)E \tag{7}$$

After applying these corrections, the energy sum $E_1^o + E_2^o$ is correctly reconstructed at 511 keV, as shown in Figure 4b.

Table 1. The mean fraction of neighbor-to-main pixel energy response, $<\epsilon>$, for the three most abundant event topologies in modules A and B, respectively.

Neighbors	Pixel Distance d (mm)	$<\epsilon_A>$	$<\epsilon_B>$
1st adjacent	3.2	0.060	0.066
1st diagonal	4.5	0.020	0.019
2nd direct	6.4	0.013	0.014

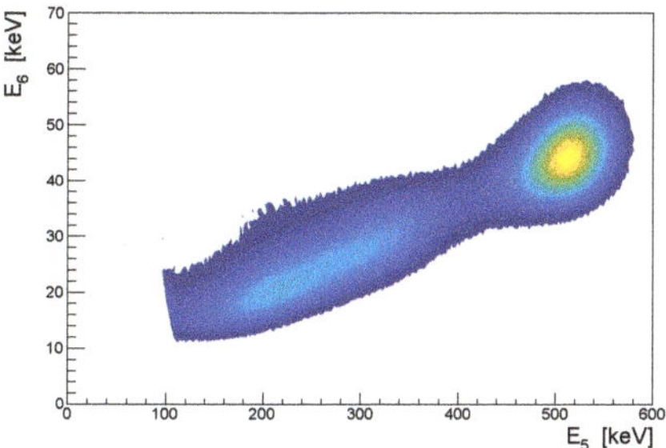

Figure 3. Observed correlation of energies between adjacent Pixel Nos. 5 and 6, attributed to light sharing.

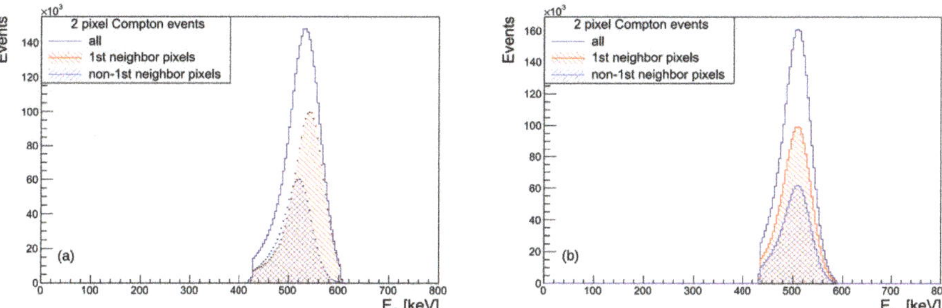

Figure 4. Observed energy of the fully absorbed 511 keV gamma: (**a**) two-pixel events before the light sharing correction; (**b**) two-pixel events after the light sharing correction.

2.5. Reconstruction of Compton Scattering Angles

The Compton events are selected by requiring two fired pixels in a module, with the sum of pixel energies, $E_1^o + E_2^o$ within $\pm 3\sigma$ of the full energy peak at 511 keV (435 keV $< E_1^o + E_2^o <$ 590 keV) and the energy of any pixel 60 keV $< E_{px}^o <$ 405 keV. In the latter condition, the lower bound of 60 keV is set conservatively to avoid possible noise contributions, and the upper bound of 405 keV is determined by the Compton edge (340 keV $\pm 3\sigma$).

In Compton scattering, the scattering angle θ is defined as the angle between the momenta vectors of the incoming and the scattered gamma $(\vec{k_0}, \vec{k'})$. The scattering angle θ does not uniquely characterize the direction of the scattered gamma; it only defines a cone on which the scattered momentum lies. Therefore, to reconstruct the momentum vector of the scattered gamma fully, one needs to determine the angle ϕ (sometimes denoted as the azimuthal angle). In our case, it is defined as the angle between the scattering plane $(\vec{k_0}, \vec{k'})$ and the horizontal plane $(\hat{x}, \hat{z})$.

2.5.1. Scattering Angle θ

In Compton scattering, the scattering angle θ is related to the energy of the scattered gamma by kinematics:

$$\theta = \mathrm{acos}\left(m_e c^2 \left(\frac{1}{E_\gamma} - \frac{1}{E'_\gamma} \right) - 1 \right) \tag{8}$$

However, one needs to determine which of the two fired pixels' energies corresponds to the recoil electron energy, E'_e, and which to the scattered photon energy, E'_γ. Specifically for $E_\gamma = 511$ keV and $\theta < 60°$, this is uniquely determined, since in this case, $E'_e < E'_\gamma$. For $\theta > 60°$, corresponding to pixel energies 171 keV $< E_{px}^o <$ 340 keV, the scattering angle determination is ambiguous, as it is not possible to directly discriminate between the forward and backward scattering, unless the detector provides depth-of-interaction information; e.g., for $\theta = 70°$, one expects the fired pixels with energies $E_e' = 203$ keV and $E_\gamma' = 308$ keV. However, scattering at $\theta = 121°$ would result in the same pixel energies, $E_e' = 308$ keV and $E_\gamma' = 203$ keV. According to Klein–Nishina relation, the ratio of the differential cross-sections for scattering at these angles is $\eta_{f/b} = \frac{d\sigma}{d\Omega}|_{\theta=70°} / \frac{d\sigma}{d\Omega}|_{\theta=121°} = 1.46$, meaning that the forward scattering is more probable.

The forward-backward ambiguity can be further suppressed by exploiting the detector design (i.e., segmentation and material). The backward scattered gammas have lower energy than the forward scattered ones, so they will have a shorter attenuation length. Let the distance traveled by a scattered gamma be $D = d/\sin\theta$, where d is the pixel distance in the x-y plane. The ratio of the forward to backward scattered gammas is then:

$$\xi_{f/b} = e^{\frac{-D(\theta_f)}{\mu(E_f)}} / e^{\frac{-D(\theta_b)}{\mu(E_b)}} \tag{9}$$

where $\mu(E)$ is the attenuation length in the material for given energy E and $D(\theta_f)$ and $D(\theta_b)$ are the pathlengths for forward and backward scattering, respectively. The ratio of the probabilities to observe forward versus backward scattering is $P_f/P_b = \eta_{f/b} \times \xi_{f/b}$. Table 2 gives an example for two angles in the region of interest ($\theta > 60°$).

In our approach, the scattering angle θ is reconstructed assuming the forward scattering, meaning the pixel with the higher energy (E_{high}) is associated with the scattered gamma, while the pixel with the lower energy (E_{low}) is associated with the recoil electron. For 511 keV gammas, this is always the case for $\theta < 60°$, but it is also justified to assume so for larger angles since the probability to observe the forward scattering is always larger than the one to observe the backward scattering, $P_f/P_b > 1$.

Table 2. Estimated ratio of forward to backward scattering probabilities for two angle combinations θ with ambiguous energy response. Attenuation lengths in the Lutetium Fine Silicate (LFS) scintillator derived from [13].

θ_f/θ_b	$\eta_{f/b}$	Pixel Distance d (mm)		$\xi_{f/b}$	P_f/P_b
80°/102°	1.20	3.2	(1st neighbors)	1.4	1.7
		6.4	(2nd neighbors)	1.9	2.3
70°/121°	1.46	3.2	(1st neighbors)	2.5	3.6
		6.4	(2nd neighbors)	6.1	8.9

The uncertainty in the determination of the scattering angle is dominated by the energy resolution of the pixels. After substituting $E_\gamma = E_{e'} + E_{\gamma'}$ in Equation (8), it follows:

$$\sigma_\theta^2 = \left(\frac{\partial\theta}{\partial E_{e'}}\sigma_{E_{e'}}\right)^2 + \left(\frac{\partial\theta}{\partial E_{\gamma'}}\sigma_{E_{\gamma'}}\right)^2 \tag{10}$$

An empirical estimate of the uncertainty is obtained as follows: we assume $E_{high} = E_{\gamma'}$ and $E_{low} = E_{e'}$; therefore, one can obtain the $\frac{\partial\theta}{\partial E_{\gamma'}}$ and $\frac{\partial\theta}{\partial E_{e'}}$ slopes from the measured distributions as in Figure 5. Further, the σ_E at lower energies can be estimated by scaling the average single pixel energy resolution at 511 keV (12.2% or $\sigma_E = 26.5$ keV) as $\sigma_E \sim \sqrt{E}$. It follows that the angular uncertainty for, e.g., $\theta = 70°$ is $\sigma_\theta = 7.9°$, and for $\theta = 80°$, it is $\sigma_\theta = 8.0°$, the latter being equivalent to $\Delta\theta = 18.8°$ (FWHM).

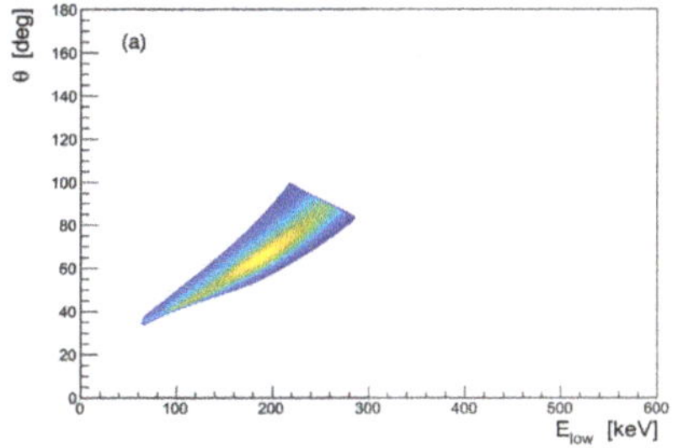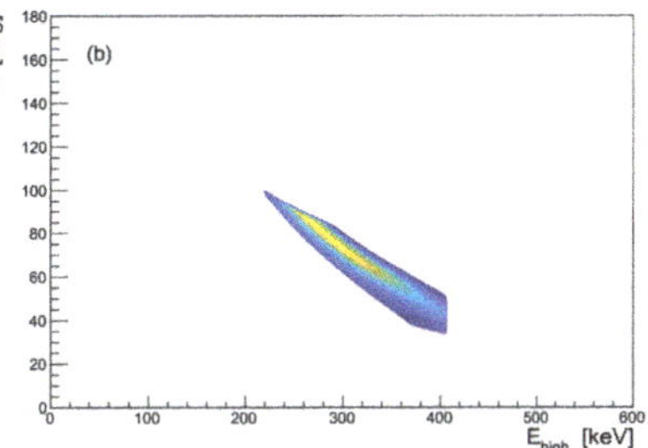

Figure 5. Reconstructed scattering angle versus response in: (**a**) lower-energy pixel; (**b**) higher-energy pixel.

2.5.2. The Angle ϕ

For gammas that Compton scatter inside the detector module, the angle ϕ is determined from the position of the fired pixels. According to Figure 1, one can write:

$$\tan\phi = \frac{y_2 - y_1}{x_2 - x_1} \tag{11}$$

where (x_1, y_1) are the coordinates of the center of the pixel where the recoil electron is detected and (x_2, y_2) are the coordinates of the center of the pixel where the scattered gamma is detected.

The uncertainty of the ϕ angle is dominated by the pixel size, i.e., by the uncertainty of the interaction position within the pixel:

$$\sigma_\phi^2 = \left(\frac{\partial\phi}{\partial x_1}\sigma_{x_1}\right)^2 + \left(\frac{\partial\phi}{\partial x_2}\sigma_{x_2}\right)^2 + \left(\frac{\partial\phi}{\partial y_1}\sigma_{y_1}\right)^2 + \left(\frac{\partial\phi}{\partial y_2}\sigma_{y_2}\right)^2 \tag{12}$$

Since all the pixels have the same dimensions, the expression for the uncertainty becomes:

$$\sigma_\phi^2 = 2\left(\frac{\partial\phi}{\partial x}\sigma_x\right)^2 + 2\left(\frac{\partial\phi}{\partial y}\sigma_y\right)^2 \tag{13}$$

The partial derivatives follow from Equation (11); hence, we get:

$$\sigma_\phi^2 = \frac{2(x^2 + y^2)}{(x^2 + y^2)^2}\sigma^2 \tag{14}$$

We can write $(x^2 + y^2) = d^2$, where d is simply the distance between the pixels (in the x-y plane). Furthermore, from x-y symmetry, it follows that $\sigma_x = \sigma_y = \sigma = a/\sqrt{12}$, where the last equality is know as the standard deviation of a uniform distribution with the width a. The expression for the uncertainty in ϕ simplifies to:

$$\sigma_\phi = \frac{1}{\sqrt{6}}\left|\frac{a}{d}\right| \tag{15}$$

where $a = 3.14$ mm, the width of the crystal pixel; e.g., for the adjacent neighbors, $\sigma_\phi = 23.0°$, and for the second neighbors, it is $\sigma_\phi = 11.5°$ or expressed as the full-width at half maximum, the angular resolutions amount $\Delta\phi = 54°$ and $\Delta\phi = 27°$, respectively.

2.5.3. Acceptance Correction

The ϕ-acceptance for the gammas that Compton scatter inside the module is not uniform, because they are more attenuated for the angles covered by distant pixel pairs. To correct this effect, we estimate the ϕ-acceptance experimentally, by plotting the normalized ϕ-distribution, ϕ^{norm}, obtained for all triggered 511 keV gammas that undergo Compton scattering in a module, as shown in Figure 6. The acceptance-corrected ϕ distribution, for any subset of measured Compton scattering events, is then obtained according to:

$$\phi(bin) = \frac{\phi^{measured}(bin)}{\phi^{norm}(bin)} \tag{16}$$

where *bin* represents a bin in the histogram.

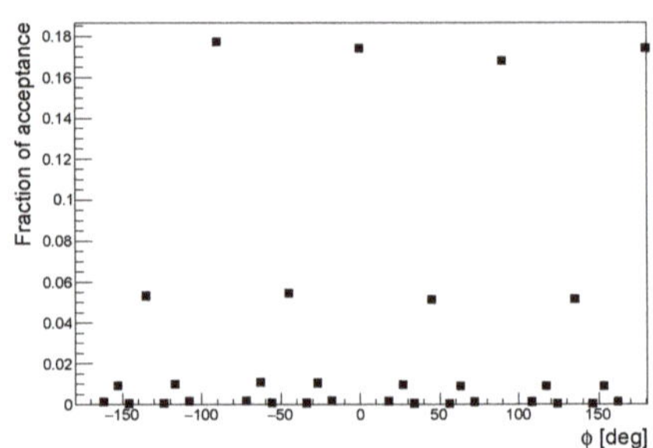

Figure 6. The normalized ϕ-acceptance for Compton scattering of 511 keV gammas in a module.

2.6. Time Reconstruction

To determine the time of the signal arrival, we apply the leading edge time pick off, described in Section 2.3. However, additional corrections are applied to improve the coincidence time resolution between detector module A and the detector module B.

2.6.1. Channel-To-Channel Correction

The coincidence time spectrum, $\Delta t_{a,b} = t_a - t_b$, is determined for each pair of channels (a, b), where $a = 0...15$ from detector A and $b = 0...15$ from detector B, using single pixel events, in which the full energy of the first gamma is deposited in a single pixel in one module and the full energy of the second gamma is deposited in a single pixel in the other module. We have observed that coincidence peaks in the time spectra have different offsets, which vary from channel to channel in the range -0.8 ns–0.6 ns, as a consequence of small channel-to-channel hardware variations (Figure 7a). However, the offsets are fixed in time and can be subtracted. After this correction, the coincidence time spectra of any channels pairs are positioned as zero, as demonstrated in Figure 7b.

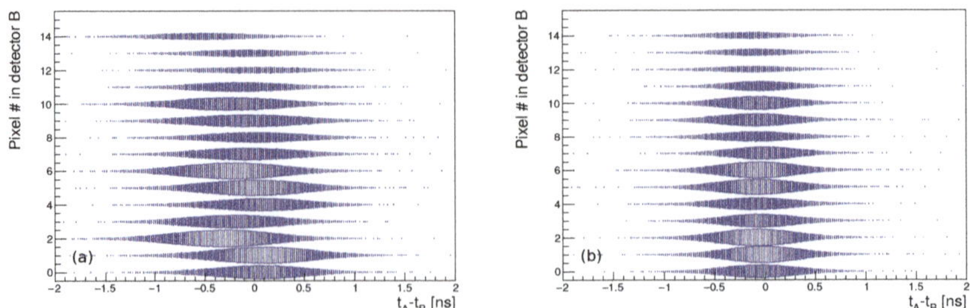

Figure 7. Channel-to-channel timing variations: (**a**) before correction; (**b**) after correction.

2.6.2. Walk Correction

It is known that the leading-edge timing method suffers from the so-called walk effect dependence of the threshold crossing time on the signal amplitude. For the coincidence time, $\Delta t = t_A - t_B$, this is translated into dependence on $E_A - E_B$, where E_A and E_B are total energies in modules A and B, respectively (Figure 8a). To correct the effect, we fit a first order polynomial $\Delta t = k\,\Delta E + l$ to data from which the slope k is determined. The corrected spectrum is obtained as:

$$\Delta t_c = \Delta t + k\,\Delta E \tag{17}$$

The $\Delta E - \Delta t$ diagram after applying this correction is shown in Figure 8b. This improves the coincidence time resolution by $\sim 4\%$.

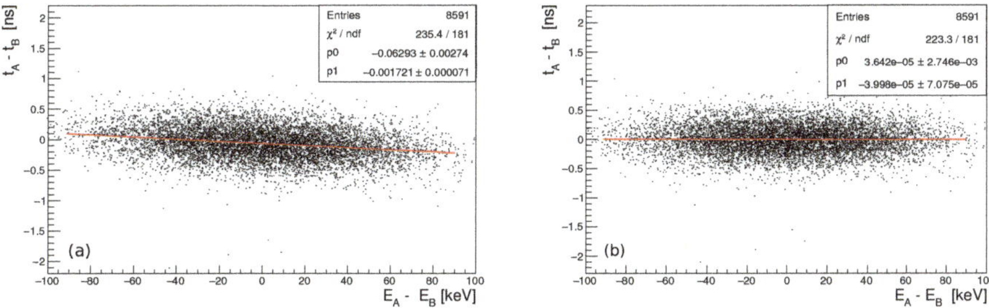

Figure 8. Coincidence time $\Delta t = t_A - t_B$ vs. energy difference $\Delta E = E_A - E_B$, for Compton events. Histogram (**a**) shows the walk effect as a clear dependence between Δt and ΔE. Histogram (**b**) is obtained after applying the walk correction.

2.6.3. Timing in Two-Pixel Events

In two-pixel Compton events, one can define the arrival time by two channels in each module. We have compared two approaches: in the first, we determine the time in each detector module as the simple average of the two pixel times:

$$t_m = \frac{t_{m,1} + t_{m,2}}{2} \tag{18}$$

where $m = A, B$ and indices 1 and 2 refer to the two pixels fired in the module.

In the other approach, we determine the time in each module as a weighted average of the two pixel times:

$$t_m = t_{m,1}\frac{E_1}{E_m} + t_{m,2}\frac{E_2}{E_m} \tag{19}$$

where $m = A, B$ and indices 1 and 2 refer to the two pixels fired in the module. As demonstrated in Figure 9, the latter method yields a better result.

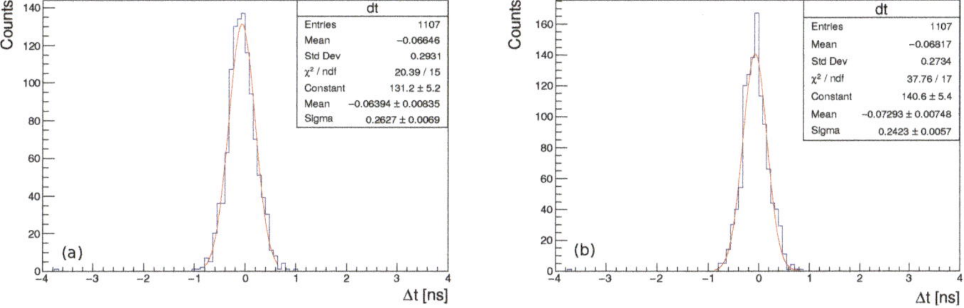

Figure 9. Coincidence time spectra for selected Compton events, after applying channel-by-channel and walk corrections. In (**a**) is shown the coincidence time spectrum calculated using the arithmetic mean. In (**b**) is the coincidence time spectrum calculated using the weighted mean.

3. Results

We investigated several aspects of detector performance in Compton scattering of 511 keV gamma particles: the ability to detect the Compton scattering events, including the reconstruction of the scattering angles (θ, ϕ), as well as the corresponding energy and angular resolutions. We also tested the timing performance of two modules in coincident detection of two 511 keV gamma particles from positron annihilation.

3.1. Reconstruction of Compton Events

To select the events where Compton scattering occurs inside a module, we require that the total deposited gamma energy, defined as the sum of the fired pixels' energies, is within the $\pm 3\sigma$ range from the peak maximum, as in Figure 10a. Generally, in those events, one or more pixels could have fired, contributing to the total energy. The relative abundances of events with different multiplicities of fired pixels are summarized in Table 3. By selecting the events with two fired pixels and full energy deposition, we select the Compton scattered gammas. The obtained energies of contributing pixels are shown in Figure 10b–d.

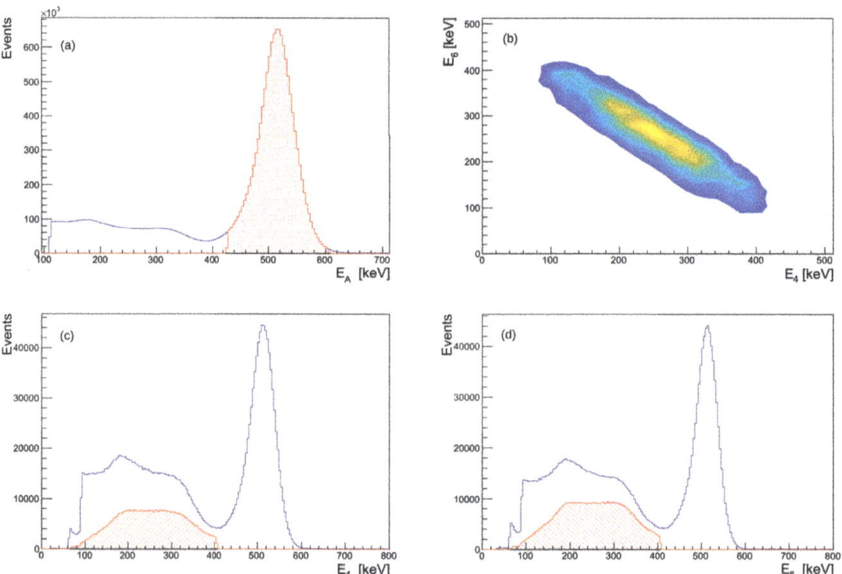

Figure 10. Compton event reconstruction: (**a**) Energy deposition in detector module A. The selected energy range corresponding to full energy deposition is shaded. (**b**) Energy in two pixels that fire in a Compton event; in this example, pixel 6 vs. pixel 4 (second neighbors). (**c**) Energy deposition in pixel 4, for all triggered events (full spectrum) and for filtered Compton events (shaded). (**d**) Energy deposition in pixel 6, for all triggered events (full spectrum) and for filtered Compton events (shaded).

Table 3. Abundance of events with different multiplicities relative to the total triggered events.

Events	Fraction of Events
Total triggered	1
Full energy deposition in the module	0.62
of which events with:	
1 pixel fired	0.46
2 pixels fired	0.15
>2 pixels fired	0.01

3.2. Energy Resolution

The Compton events may have different event topologies in a detector module reflected in different mean inter-pixel distances, d, ranging from 3.2 mm for the adjacent pixels to 13.6 mm for the farthest pixel pairs (see Figure 1). A specific event topology can be chosen by selecting the corresponding pixel distance. The total gamma energy, reconstructed as the sum of the fired pixels' energies, is shown in Figure 4a. One observes the shift of the position of the full energy peak depending on event topology. Most notably, the peak maximum is shifted upwards ~30 keV for the events where the fired pixels are adjacent neighbors, while much lower shifts (<10 keV) are observed when the fired pixels are more distant neighbors. The observed shifts had been attributed to light sharing between the pixels [11]. A procedure to correct this effect and reconstruct the real energy depositions is presented in Section 2.4.2. Upon applying it, the gamma energies are correctly reconstructed for all event topologies, as demonstrated in Figure 4b.

The energy resolution at 511 keV, defined as the full width at half maximum of the energy sum peak, was 12.4% ± 0.1% for the Compton events, where the fired pixels are adjacent neighbors, and it was 12.0% ± 0.2% for the case where the fired pixels are farther neighbors. This is consistent with the observed mean energy resolution of single pixels, which was 12.2% ± 0.7%.

3.3. Angular Resolution

The scattering angle θ was reconstructed from the measured pixel energies using Compton scattering kinematics (see Section 2.5 for details). The reconstructed angles were in the range of $40° < \theta < 90°$, as shown in Figure 11a, where the lower limit ($\theta = 40°$) was set by the pixel energy threshold, while the upper limit was determined by the reconstruction algorithm, which relies on suppression of the observed scattering at angles $\theta > 90°$, due to a lower cross-section and short attenuation length. The angular resolution was limited by the energy resolution of the pixels, and it was approximately constant throughout the acceptance, being $\Delta\theta \simeq 18.8°$ (FWHM).

The angle ϕ was reconstructed from the relative positions of the two fired pixels. In this case, the module had a full 2π acceptance; however, it was non-uniform, because the scattered gammas were more attenuated at the angles covered by more distant pixel pairs, as shown in Figure 6. This can be corrected, as described in Section 2.5.3. Figure 11b shows an example of the acceptance-corrected ϕ distribution for coincident Compton events in two modules. The angular resolution in ϕ depends on the distance of the fired pixels, ranging from $\Delta\phi = 12.7°$ (FWHM) for the largest pixel distance d to $\Delta\phi = 54°$ (FWHM) when the fired pixels are adjacent neighbors.

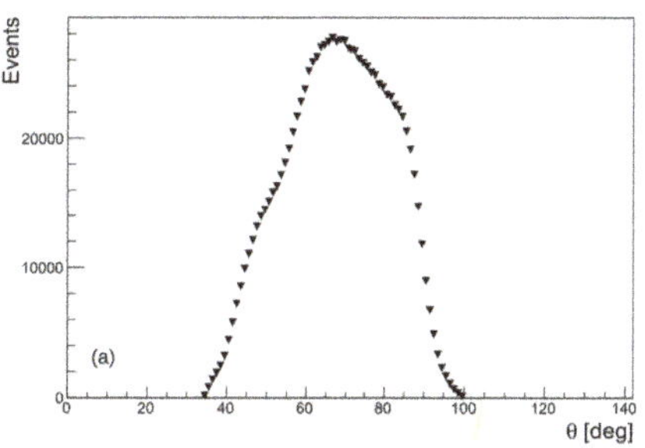
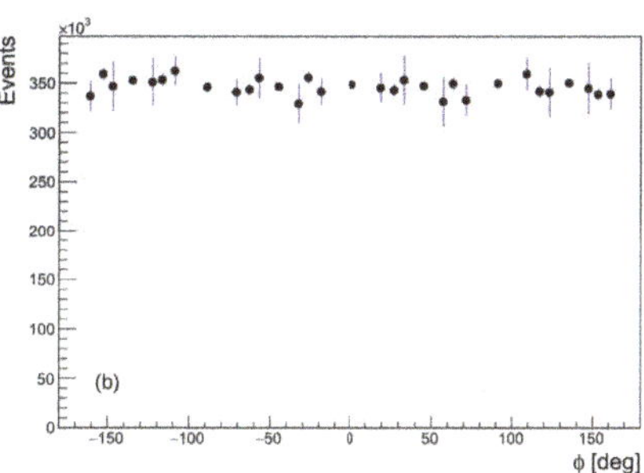

Figure 11. Example of reconstructed angles in Compton scattering: (**a**) the scattering angle θ; (**b**) the angle ϕ (acceptance corrected).

3.4. Coincidence Time Resolution

The coincidence time resolution (CTR) of the system of two modules has been evaluated for events where Compton scattering occurs in both modules. In order to minimize the influence of events with incomplete energy absorption in a module, we required the energy in each module to be ±2σ

from the full energy peak. The arrival time of each signal was reconstructed by the leading edge method, on top of which we applied walk correction and pixel-to-pixel offset correction, described in detail in Section 2.6.1.

In Compton events, two fired pixels in each module can determine time, and we have examined two approaches to combine this information. In the first, the time in a module was determined as the arithmetic mean of the times reconstructed from pixels, while the second approach used the weighted mean, with weights corresponding to energy deposited in each pixel. The time difference between the modules $\Delta t = t_A - t_B$ is plotted in Figure 9. The first method resulted in $\Delta t = 0.61 \pm 0.02$ ns (FWHM), while the second method yielded $\Delta t = 0.56 \pm 0.02$ ns (FWHM). The latter is consistent with the result $\Delta t = 0.54 \pm 0.02$ ns (FWHM) obtained for single-pixel (photo-electric absorption) events, under the same operating conditions.

4. Discussion and Conclusions

We have presented the performance of the single-layer scintillator pixel detectors, demonstrating that it is possible to detect Compton scattered gamma particles with energy and timing resolutions comparable to those achieved in photo-electric absorption, while it is also possible to reconstruct the direction of the scattered gammas. The detection and full reconstruction of Compton scattering has been of interest in medical imaging, such as PET, where measurement of polarization correlations of annihilation quanta had been studied to improve sensitivity. This work proves the concept of the full Compton scattering reconstruction in modules with only one readout layer. Compared to two-layer Compton detectors, the number of electronic channels is reduced by a half; hence, the application of single-layer modules could have the potential to significantly improve the cost efficiency for larger devices with Compton detection capabilities.

Author Contributions: Conceptualization, M.M. and D.B.; data curation, M.M. and L.P.; formal analysis, M.M.; funding acquisition, M.M. and D.B.; investigation, M.M. and L.P.; methodology, M.M.; project administration, M.M. and D.B.; resources, M.M.; software, M.M. and L.P.; supervision, M.M.; visualization, M.M.; writing, original draft, M.M.

Funding: This work has been supported in part by the Croatian Agency for Small and Medium Enterprises, Innovations and Investments (HAMAG-BICRO), Proof-of-Concept Programme, Project POC6_1_211, and in part by the Croatian Science Foundation under Project 8570.

Acknowledgments: We acknowledge inputs by the European Cooperation for Science and Technology Action TD1401: Fast Advanced Scintillation Timing (http://cern.ch/fast-cost).

Conflicts of Interest: The authors declare no conflict of interest.

References

1. Kuncic, Z.; McNamara, A.; Wu, K.; Boardman, D. Polarization enhanced X-ray imaging for biomedicine. *Nucl. Instrum. Methods Phys. Res. A* **2011**, *648*, S208–S210. [CrossRef]
2. McNamara, A.; Toghyani, M.; Gillam, J.; Wu, K.; Kuncic, Z. Towards optimal imaging with PET: An in silico feasibility study. *Phys. Med. Biol.* **2014**, *59*, 7587–7600. [CrossRef] [PubMed]
3. Toghyani, M.; Gillam, J.; McNamara, A.; Kuncic, Z. Polarisation-based coincidence event discrimination: An in silico study towards a feasible scheme for Compton-PET. *Phys. Med. Biol.* **2016**, *61*, 5803–5817. [CrossRef] [PubMed]
4. Mitani, T.; Tanaka, T.; Nakazawa, K.; Takahashi, T.; Takashima, T.; Tajima, H.; Nakamura, H.; Nomachi, M.; Nakamoto, T.; Fukazawa, Y. A prototype Si/CdTe Compton camera and the polarization measurement. *IEEE Trans. Nucl. Sci.* **2004**, *51*, 2432–2437. [CrossRef]
5. Takeda, S.; Odaka, H.; Katsuta, J.; Ishikawa, S.; Sugimoto, S.; Koseki, Y.; Watanabe, S.; Sato, G.; Kokubun, M.; Takahashi, T.; et al. Polarimetric performance of Si/CdTe semiconductor Compton camera. *Nucl. Instrum. Methods Phys. Res. A* **2010**, *622*, 619–627. [CrossRef]
6. Yonetoku, D.; Murakami, T.; Gunji, S.; Mihara, T.; Sakashita, T.; Morihara, Y.; Kikuchi, Y.; Takahashi, T.; Fujimoto, H.; Toukairin, N.; et al. Gamma-Ray Burst Polarimeter-GAP-aboard the Small Solar Power Sail Demonstrator IKAROS. *Publ. Astron. Soc. Jap.* **2011**, *63*, 625–638. [CrossRef]

7. Shimazoe, K.; Yoshino, M.; Ohshima, Y.; Uenomachi, M.; Oogane, K.; Orita, T.; Takahashi, H.; Kamada, K.; Yoshikawa, A.; Takahashi, M. Development of simultaneous PET and Compton imaging using GAGG-SiPM based pixel detectors. *Nucl. Instrum. Methods Phys. Res. A* **2018**. [CrossRef]
8. Uenomachi, M.; Mizumachi, Y.; Yoshihara, Y.; Takahashi, T.; Shimazoe, K.; Yabu, G.; Yoneda, H.; Watanabe, S.; Takeda, S.; Orita, T.; et al. Double photon emission coincidence imaging with GAGG-SiPM Compton camera. *Nucl. Instrum. Methods Phys. Res. A* **2018**. [CrossRef]
9. Spillmann, U.; Bräuning, H.; Hess, S.; Beyer, H.; Stöhker, T.; Dousse, J.C.; Protic, D.; Krings, T. Performance of a Ge-microstrip imaging detector and polarimeter. *Rev. Sci. Instrum.* **2008**, *79*, 083101. [CrossRef] [PubMed]
10. Bloser, P.F.; Legere, J.S.; McConnell, M.L.; Macri, J.R.; Bancroft, C.M.; Connor, T.P.; Ryan, J.M. Calibration of the Gamma-RAy Polarimeter Experiment (GRAPE) at a Polarized Hard X-Ray Beam. *Nucl. Instrum. Methods Phys. Res. A* **2009**, *600*, 424–433. [CrossRef]
11. Makek, M.; Bosnar, D.; Gačić, V.; Pavelić, L.; Šenjug, P.; Žugec, P. Performance of scintillation pixel detectors with MPPC read-out and digital signal processing. *Acta Phys. Pol. B* **2017**, *48*, 1721–1726. [CrossRef]
12. Gundacker, S.; Auffray, E.; Pauwels, K.; Lecoq, P. Measurement of intrinsic rise times for various L(Y)SO and LuAG scintillators with a general study of prompt photons to achieve 10 ps in TOF-PET. *Phys. Med. Biol.* **2016**, *61*, 2802–2837. [CrossRef] [PubMed]
13. Berger, M.; Hubbell, J.; Seltzer, S.; Chang, J.; Coursey, J.; Sukumar, R.; Zucker, D.; Olsen, K. XCOM: Photon Cross Section Database (Version 1.5). Available online: http://physics.nist.gov/xcom (accessed on 30 January 2019).

Article

High Precision Test of the Pauli Exclusion Principle for Electrons

Kristian Piscicchia [1,2,*], Aidin Amirkhani [3], Sergio Bartalucci [2], Sergio Bertolucci [4], Massimiliano Bazzi [2], Mario Bragadireanu [2,5], Michael Cargnelli [6], Alberto Clozza [2], Catalina Curceanu [1,2,5], Raffaele Del Grande [2], Luca De Paolis [2], Jean-Pierre Egger [7], Carlo Fiorini [3], Carlo Guaraldo [2], Mihail Iliescu [2], Matthias Laubenstein [8], Johann Marton [6], Marco Miliucci [2], Edoardo Milotti [9], Andreas Pichler [6], Dorel Pietreanu [2,5], Alessandro Scordo [2], Hexi Shi [10], Diana Laura Sirghi [2,5], Florin Sirghi [2,5], Laura Sperandio [2], Oton Vazquez Doce [11] and Johann Zmeskal [6]

[1] Centro Fermi—Museo Storico della Fisica e Centro Studi e Ricerche "Enrico Fermi", 00184 Rome, Italy; Catalina.Curceanu@lnf.infn.it
[2] Laboratori Nazionali di Frascati, INFN, 00044 Frascati, Italy; Sergio.bartalucci@lnf.infn.it (S.B.); massimiliano.bazzi@lnf.infn.it (M.B.); bragadireanu.mario@lnf.infn.it (M.B.); alberto.clozza@lnf.infn.it (A.C.); raffaele.delgrande@lnf.infn.it (R.D.G.); Luca.DePaolis@lnf.infn.it (L.D.P.); guaraldo@lnf.infn.it (C.G.); mihai.iliescu@lnf.infn.it (M.I.); marco.miliucci@lnf.infn.it (M.M.); dorel.pietreanu@lnf.infn.it (D.P.); scordo@lnf.infn.it (A.S.); sirghi@lnf.infn.it (D.L.S.); fsirghi@lnf.infn.it (F.S.); Laura.Sperandio@lnf.infn.it (L.S.)
[3] Politecnico di Milano, Dipartimento di Elettronica, Informazione e Bioingegneria and INFN Sezione di Milano, 20133 Milano, Italy; aidin.amirkhani@polimi.it (A.A.); carlo.fiorini@polimi.it (C.F.)
[4] Dipartimento di Fisica e Astronomia, Università di Bologna, 40126 Bologna, Italy; Sergio.Bertolucci@lnf.infn.it
[5] IFIN-HH, Institutul National pentru Fizica si Inginerie Nucleara Horia Hulubei, Magurele 077125, Romania
[6] Stefan-Meyer-Institute for Subatomic Physics, Austrian Academy of Science, 1090 Vienna, Austria; michael.cargnelli@oeaw.ac.at (M.C.); johann.marton@oeaw.ac.at (J.M.); Andreas.Pichler@oeaw.ac.at (A.P.); johann.zmeskal@oeaw.ac.at (J.Z.)
[7] Institut de Physique, Université de Neuchâtel, CH-2000 Neuchâtel, Switzerland; jean-pierre.egger@net2000.ch
[8] Laboratori Nazionali del Gran Sasso, INFN, 67100 Assergi, Italy; matthias.laubenstein@lngs.infn.it
[9] Dipartimento di Fisica, Università di Trieste and INFN-Sezione di Trieste, 34127 Trieste, Italy; milotti@ts.infn.it
[10] Institut für Hochenergiephysik der Österreichischen Akademie der Wissenschaften, 1050 Vienna, Austria; shihexi@gmail.com
[11] Excellence Cluster Universe, Technische Universität München, D-85748 Garching, Germany; otonvazquezdoce@gmail.com
[*] Correspondence: kristian.piscicchia@lnf.infn.it

Received: 6 March 2019; Accepted: 23 April 2019; Published: 2 May 2019

Abstract: The VIP-2 experiment aims to perform high precision tests of the Pauli Exclusion Principle for electrons. The method consists in circulating a continuous current in a copper strip, searching for the X radiation emission due to a prohibited transition (from the 2p level to the 1s level of copper when this is already occupied by two electrons). VIP already set the best limit on the PEP violation probability for electrons $\frac{1}{2}\beta^2 < 4.7 \times 10^{-29}$, the goal of the upgraded VIP-2 (VIolation of the Pauli Exclusion Principle-2) experiment is to improve this result of two orders of magnitude at least. The experimental apparatus and the results of the analysis of a first set of collected data will be presented.

Keywords: Pauli exclusion principle; quantum foundations; X-ray spectroscopy; underground experiment

---

1. Introduction

The VIP collaboration is performing high precision tests of the Pauli Exclusion Principle (PEP) for electrons, in the extremely low cosmic background environment of the Underground Gran Sasso Laboratories (LNGS) of INFN (Italy). According to the PEP a system can not hold two (or more) fermions with all quantum numbers identical. PEP stands as one of the fundamental and most solid cornerstones of modern physics, its validity explaining a plenty of phenomena such as the structure of atoms. The PEP was originally formulated for the electrons (see Ref. [1]) and was later extended to all the fermions by the spin-statistics theorem, which can only be demonstrated within Quantum Field Theory (QFT). According to the spin-statistics connection the quantum states of identical particles are necessarily either symmetric (for bosons) or antisymmetric (for fermions) with respect to their permutation. Extensions of the QFT admit, however, spin-statistics violations, hence experimental evidence of even a tiny violation of the PEP would be an indication of physics beyond the Standard Model.

VIP (see Refs. [2–4]) greatly improved an experimental technique conceived by Ramberg and Snow (see Ref. [5]) which consists in circulating a DC current in a copper conductor and search for the X-rays signature of PEP-violating K_α transitions ($2p \rightarrow 1s$ in Cu when the $1s$ level is already occupied by two electrons). As a consequence of the shielding effect of the two electrons in the ground state, the K_α violating transition is shifted of about 300 eV with respect to the standard line and is then distinguishable in precision spectroscopic measurements. Such experimental procedure aims to evidence an anomalous behaviour of the newly injected electrons which never had before the possibility to perform the searched violating K_α transition in the target Cu atoms. In this sense VIP strictly fulfills the Messiah-Greenberg superselection rule [6] which excludes transitions between different symmetry states in a given system. Considering open systems is then a crucial feature in order to consistently test a violation of the PEP, whose probability is usually quantified by means of the $\beta^2/2$ parameter [7,8].

In what follows the upgraded VIP-2 experimental apparatus (see Refs. [9,10]) will be presented, and the analysis of a first set of data (collected in 2016) will be described. As will be shown VIP-2 already improved the upper limit imposed by VIP on $\beta^2/2$ (after three years of data taking), which represents the best limit ever on the PEP violation probability for electrons. The final goal of VIP-2 (which is presently acquiring data) is to either further improve the limit of two orders of magnitude, or to measure a signal of PEP violation.

2. The VIP-2 Experimental Apparatus

VIP-2 is the upgraded version of the VIP experiment and aims to improve the result obtained by VIP of two orders of magnitude at least. VIP set the best limit on the PEP violation probability for electrons $\frac{1}{2}\beta^2 < 4.7 \times 10^{-29}$ [2] exploiting the experimental technique which was pioneered by Ramberg and Snow. The VIP experimental setup made use of Charge Coupled Devices (CCDs) as the X-ray detectors; CCDs were characterised by a Full Width at Half Maximum (FWHM) of 320 eV at 8 keV, corresponding to the definition of the Region Of Interest (ROI) where the signature of anomalous X-ray transitions is searched for. Moreover, VIP was operated in the extremely low cosmic background environment of the Underground Gran Sasso Laboratories (LNGS) of INFN.

The goal of VIP-2 will be achieved by implementing many improvements in the experimental apparatus. The core components of the VIP-2 setup are illustrated in Figure 1. The new layout of the copper target consists of two strips of copper (with a thickness of 50 µm, and a surface of 9 cm $\times$ 2 cm) the new geometry results in a higher acceptance for the X-ray detection. The heat due to the dissipation in copper would lead to a significant temperature rise in the strip. In order to avoid this effect a cooling pad (cooled down by a closed chiller circuit) is placed in between the two strips. This also allows to enhance the DC current circulating on the strips to 100 A (instead of the 40 A in VIP) thus increasing the candidate event pool for the anomalous X-rays. The CCDs were replaced by Silicon Drift Detectors (SDDs) as X-ray detectors, with a better energy resolution (190 eV FWHM at 8 keV). The data presented in this work were acquired by means of two arrays of 1×3 SDDs surrounding the copper target, each array with 3 cm^2 of effective surface. The SDDs were cooled down to $-170\,^\circ$C with circulating liquid argon in a closed cooling line. With a current of 100 A circulating in the strips, their temperature rises up by about 20 $^\circ$C, inducing a temperature rise at the SDDs of about 1 K, which does not significantly alters the SDDs performances.

The timing capability of the SDDs also enables to introduce an active shielding system. This veto system is made of 32 plastic scintillator bars (250 mm $\times$ 38 mm $\times$ 40 mm bar) surrounding the SDDs and serves to remove the background originating from the high energy charged particles that are not shielded by the rocks of the Gran Sasso mountains. The light output of each scintillator is read out by two silicon photomultipliers (SiPMs) coupled to each end of the bars.

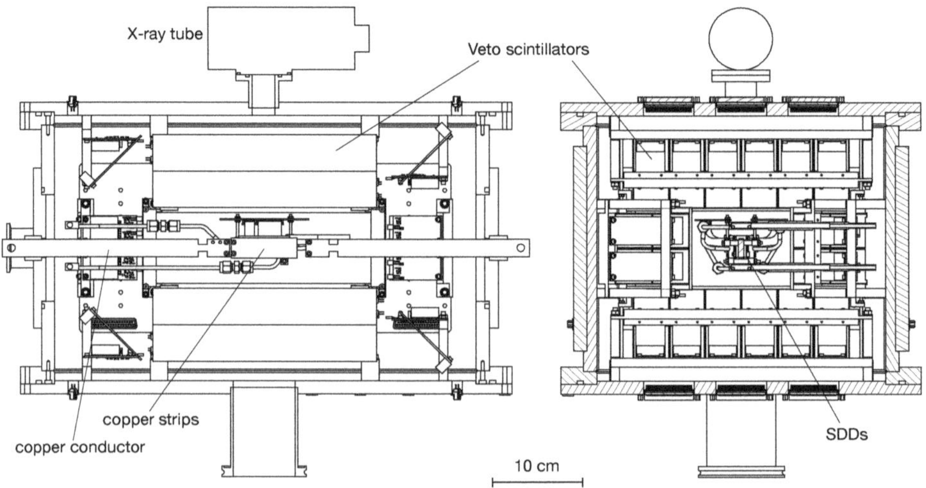

Figure 1. The side views of the design of the core components of the VIP-2 setup, including the SDDs as the X-ray detector, the scintillators as active shielding with silicon photomultiplier readout.

All the detectors and the front end preamplifier electronics are mounted inside the vacuum chamber which is kept at 10^{-5} mbar during operation.

In order to perform quick energy calibration and SDDs resolution measurements an X-ray tube on top of the setup irradiates Zirconium and Titanium foils, to produce fluorescence reference lines. A Kapton window in the vacuum chamber and an opening solid angle in the upper scintillator bars, allow to collect in one hour enough statistics for the SDDs performance monitoring. A secondary energy calibration method of the SDDs is performed by means of a weakly radioactive Fe-55 source, with a 25 µm thick Titanium foil attached on top, mounted together inside an aluminum holder. The six SDDs have an overall 2 Hz trigger rate, accumulating events of fluorescence X-rays from titanium and manganese to calibrate the digitized channel into energy scale.

The VIP-2 experimental apparatus was transported and mounted in the LNGS at the end of 2015. Following a period of tuning and optimization a first campaign of data taking started from October 2016 with the complete detector system (except the passive shielding). A total amount of 34 days of data with a 100 A DC current and 28 days without current were collected until the end of the year 2016. In the next section, the analysis of this data set and the obtained result are shown.

The VIP-2 setup was further upgraded during 2018: new copper targets were realised, the SDDs arrays were replaced with two arrays 2×8 for a total of 32 SDDs and the passive shielding was mounted. The passive shielding, which is made of two layers of lead and copper blocks, will kill most of the background due to environmental gamma radiation. The final configuration of the VIP-2 setup, which is presently taking data, is shown in Figure 2. The energy calibrated spectra corresponding to an equal data collection period of 39 days during 2018, with and without current, are shown in Figure 3 left and right respectively. The data analysis, performed with a similar procedure to that described in Section 3, is presently ongoing on this data set.

More details on the VIP-2 experimental apparatus, the trigger logic, data acquisition and slow control can be found in Ref. [9].

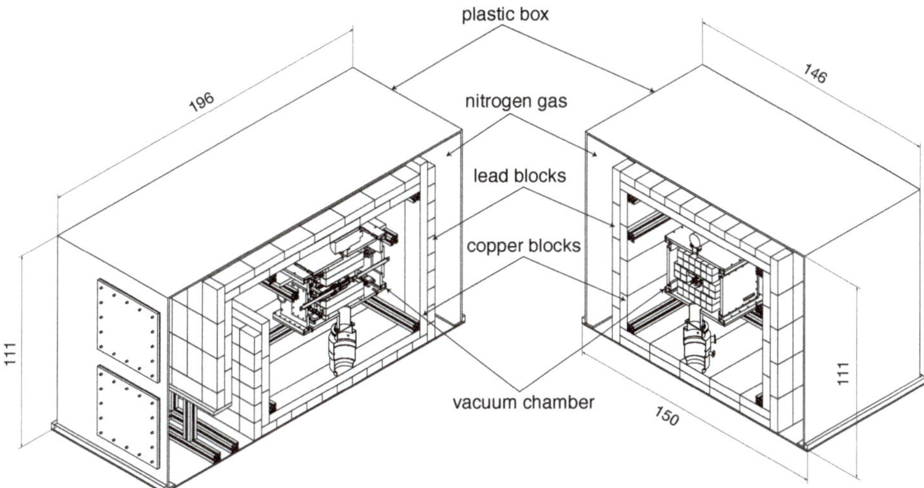

Figure 2. Perspective views of the VIP-2 apparatus with passive shielding, with the dimensions in cm. Nitrogen gas with a slight over pressure with respect to the external air will be circulated inside a plastic box in order to reduce the radon contamination.

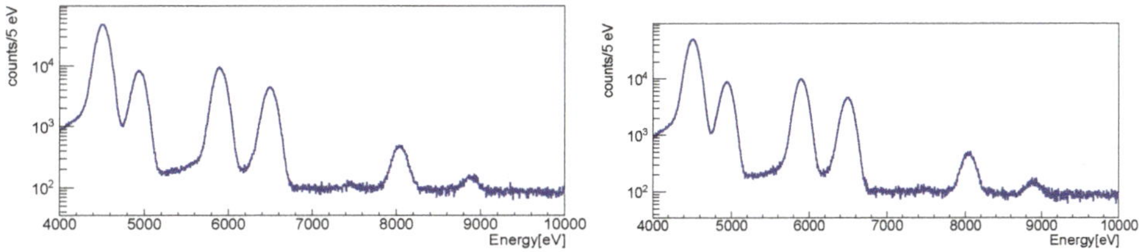

Figure 3. Energy calibrated spectra corresponding to 39 days of data taking without current (**left**) and 39 days with 100 A DC current (**right**) collected during 2018.

3. Data Analysis

In order to put in evidence an eventual signal of PEP violating K_α transitions a simultaneous fit was performed of the two spectra collected with and without current; the spectra and the obtained fit result are shown in Figure 4. The fit was performed by minimising a global Chi-square function, which is obtained as the product of the likelihoods corresponding to the two spectra, assuming the measurement errors to be distributed according to Gaussians. The fit proceeds in two steps: as first (see Figure 4a) a wide energy range is used (from 3.5 keV to 11 keV) in order to exploit the high statistics titanium and manganese lines to determine the Fano Factor and the Constant Noise (an energy independent contribution to the energy resolution). The parameters obtained from this pre-fit are then used as an input for the second fit in the range from 7 keV to 11 keV (see Figure 4b top), from which the shape of the continuous background near the interesting transition is better determined. The fit parameters accounting for the detector energy resolution, the shape of the continuous background, the shape of the fluorescence peaks, are common for the spectra with and without current. The parameters representing the intensities of the fluorescence peaks and of the continuous background are separately defined. For the current on spectrum an additional Gaussian component was introduced representing the eventual PEP violating K_α transition line, the centre of the line was set at 7746.73 eV (see Refs. [9]). In Figure 4b bottom the residuals from the second fit are shown for the two spectra. The Chi-square minimisation was performed using the MINUIT package of the CERN ROOT software framework [11]. From the fit the number of candidate PEP violating events, contributing to the K_α violating transitions, is obtained, together with the corresponding statistical error:

$$N_X = 54 \pm 67 \ (statistical).\tag{1}$$

By analogy with the original limit estimated by Ramberg and Snow in Ref. [5] N_X can be related to the PEP violation probability $\frac{1}{2}\beta^2$ as follows:

$$N_X \geq \frac{1}{2}\beta^2 \cdot N_{\text{new}} \cdot \frac{1}{10} \cdot N_{\text{int}} \cdot \epsilon.\tag{2}$$

In Equation (2) $N_{\text{new}} = (1/e)\int_{\Delta t} I(t)dt$ is the number of current electrons injected in the copper target over the acquisition time period (with current) Δt, the factor $1/10$ accounts for the capture probability (per electron-atom scattering) into the $2p$ state (see Ref. [12]), $N_{\text{int}} = D/\mu$ is the minimum number of electron-atom scatterings, where D is the effective length of the copper strip and μ the scattering length for conduction electrons in the copper strip, to conclude $\epsilon = 1.8\%$ is the detection efficiency factor, obtained by means of a Monte Carlo (MC) simulation (as described in [9]). By substituting $\mu = 3.9 \times 10^{-6}$ cm, $e = 1.602 \times 10^{-19}$ C, $I = 100$ A, and the effective length of the copper strip $D = 7.1$ cm (the same used in the MC simulation), using the three sigma upper bound of $3 \cdot \Delta N_X = 201$ to give a 99.7% C.L., the following upper limit is obtained for the PEP violation probability:

$$\frac{\beta^2}{2} \leq 3.4 \cdot 10^{-29}.\tag{3}$$

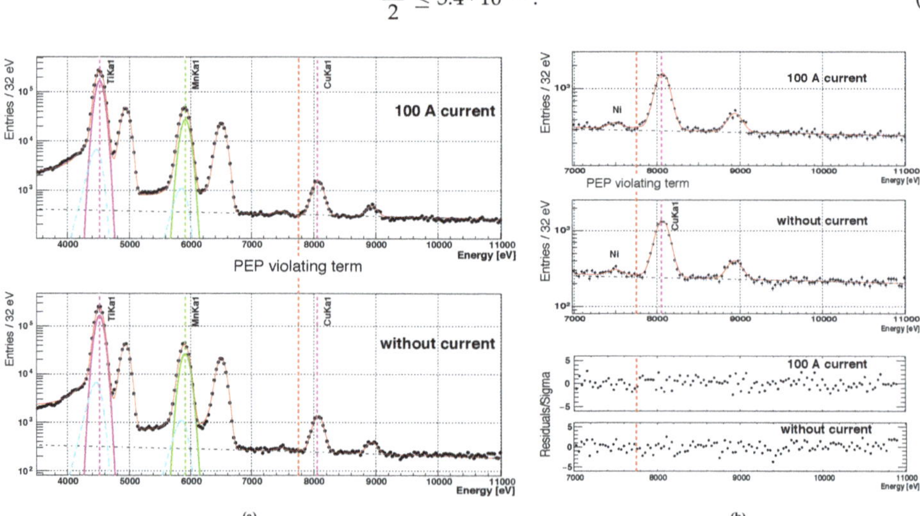

Figure 4. A global chi-square function was used to fit simultaneously the spectra with and without 100 A current applied to the copper conductor. The energy position for the expected PEP violating events is about 300 eV below the normal copper $K_{\alpha1}$ transition. The Gaussian function and the tail part of the $K_{\alpha1}$ components and the continuous background from the fit result are also plotted. (**a**): the fit to the wide energy range from 3.5 keV to 11 keV; (**b**): the fit and its residual for the 7 keV to 11 keV range where there is no background coming from the calibration source. See the main text for details.

4. Discussion and Perspectives

In analogy with the analysis performed by Ramberg and Snow, the limit obtained in Equation (3) assumes a very simple straight path of the electrons across the Cu target strip. As a consequence, the scattering length is used in order to estimate the number of electron capture processes. In Ref. [13] it is argued that scatterings are not actually related to the atoms themselves, but depend on impurities, lattice imperfections and on phonons. For this reason the mean time between close electron-atom encounters is instead evaluated in Ref. [13], which is found to be $3.5 \cdot 10^{-17}$ s (instead of the much longer average scattering time 2.5×10^{-14}). Considered the traversal time of the copper target (which is estimated in [13] to amount to 10 s for the setup described in Section 2) an improved limit is obtained on the PEP violation probability: $\frac{\beta^2}{2} \leq 2.6 \cdot 10^{-40}$.

The analysis presented in Ref. [13] for the complex random walk which electron undergo in crossing the copper target material is mostly classical. We are presently working to extend the calculation to the quantum domain.

Author Contributions: Conceptualization, S.B. (Sergio Bertolucci), M.B. (Massimiliano Bazzi), A.C., C.C., J.-P.E., C.G., M.I., M.L., J.M., E.M., A.P., D.P., J.Z.; software, K.P., M.C., R.D.G., L.S., D.P., M.S., O.V.D., D.S., H.S.; formal analysis, K.P., H.S., A.P., D.P., R.D.G., M.M., L.D.P.; data curation A.A., C.F., M.C., S.B. (Sergio Bartalucci), M.B. (Mario Bragadireanu), A.C., C.C., L.D.P., M.I., M.L., J.M., M.M., A.P., D.P., A.S., D.L.S., F.S., H.S., J.Z.; writing and editing, K.P., C.C., E.M.; supervision, C.C.

Funding: "This research was partially funded by Centro Fermi—Museo Storico della Fisica e Centro Studi e Ricerche "Enrico Fermi" (Open Problems in Quantum Mechanics project) and by the Austrian Science Foundation (FWF) with the grants P25529-N20, project P 30635-N36 and W1252-N27".

Acknowledgments: We thank Herbert Schneider, Leopold Stohwasser, and Doris Pristauz-Telsnigg from Stefan-Meyer-Institut for their fundamental contribution in designing and building the VIP2 setup. We acknowledge the very important assistance of the INFN-LNGS laboratory. We acknowledge the support of the Centro Fermi—Museo Storico della Fisica e Centro Studi e Ricerche "Enrico Fermi" (Open Problems in Quantum Mechanics project), the support from the EU COST Action CA 15220 and of the EU FET project TEQ (grant agreement 766900) is gratefully acknowledged. We thank the Austrian Science Foundation (FWF) which supports the VIP2 project with the grants P25529-N20, project P 30635-N36 and W1252-N27 (doctoral college particles and interactions). Furthermore, these studies were made possible through the support of a grant from the Foundational Questions Institute, FOXi (and a grant from the John Templeton Foundation (ID 58158). The opinions expressed in this publication are those of the authors and do not necessarily respect the views of the John Templeton Foundation.

References

1. Pauli, W. Über den Zusammenhang des Abschlusses der Elektronengruppen im Atom mit der Komplexstruktur der Spektren. *Z. Phys.* **1925**, *31*, 765. [CrossRef]
2. Petrascu, C.C.; Bartalucci, S.; Bertolucci, S.; Bragadireanu, M.; Cargnelli, M.; Di Matteo, S.; Egger, J.-P.; Guaraldo, C.; Iliescu, M.; Ishiwatari, T. Experimental tests of quantum mechanics—Pauli exclusion principle violation (the VIP experiment) and future perspective. *J. Phys. Conf. Ser.* **2011**, *306*, 012036. [CrossRef]
3. Bartalucci, S.; Bertolucci, S.; Bragadireanu, M.; Cargnelli, M.; Curceanu, C.; Di Matteo, S.; Egger, J.-P.; Guaraldo, C.; Iliescu, M.; Ishiwatari, T. The VIP experimental limit on the Pauli exclusion principle violation by electrons. *Found. Phys.* **2010**, *40*, 765–775. [CrossRef]
4. Sperandio, L. New Experimental Limit on the Pauli Exclusion Principle Violation by Electrons From the VIP Experiment. Ph.D. Thesis, Tor Vergata University, Rome, Italy, 2008.
5. Ramberg, E.; Snow, G.A. Experimental limit on a small violation of the Pauli principle. *Phys. Lett. B* **1990**, *238*, 438–441. [CrossRef]
6. Messiah, A.; Greenberg, O. Symmetrization postulate and its experimental foundation. *Phys. Rev.* **1964**, *136*, B248. [CrossRef]
7. Ignatiev, A.Y.; Kuzmin, V. *Quarks '86: Proceedings of the Seminar, Tbilisi, USSR, 15–17 April 1986*; Tavkhelidze, A.N., Matveev, V.A., Pivovarov, A.A., Tkachev, I.I., Eds.; VNU Science Press BV: Utrecht, The Netherlands, 1987; pp. 263–268.
8. Ignatiev, A.Y. X rays test the Pauli exclusion principle. *Radiat. Phys. Chem.* **2006**, *75*, 2090–2096. [CrossRef]

9.   Shi, H.; Milotti, E.; Bartalucci, S.; Bazzi, M.; Bertolucci, S.; Bragadireanu, A.; Cargnelli, M.; Clozza, A.; De Paolis, L.; Di Matteo, S.; et al. Experimental search for the violation of Pauli exclusion principle. *Eur. Phys. J. C* **2018**, *78*, 319. [CrossRef] [PubMed]

10. Curceanu, C.; Shi, H.; Bartalucci, S.; Bertolucci, S.; Bazzi, M.; Berucci, C.; Bragadireanu, M.; Cargnelli, M.; Clozza, A.; De Paolis, L.; et al. Test of the Pauli Exclusion Principle in the VIP-2 underground experiment. *Entropy* **2017**, *19*, 300. [CrossRef]

11. Brun, R.; Rademakers, F. ROOT—An object oriented data analysis framework. *Nucl. Instrum. Methods Phys. Res. A* **1997**, *389*, 81–86. [CrossRef]

12. Bartalucci, S.; Bertolucci, S.; Bragadireanu, M.; Cargnelli, M.; Catitti, M.; Curceanu, C.; Di Matteo, S.; Egger, J.-P.; Guaraldo, C.; Iliescu, M. New experimental limit on the Pauli exclusion principle violation by electrons. *Phys. Lett. B* **2006**, *641*, 18–22. [CrossRef]

13. Milotti, E.; Bartalucci, S.; Bertolucci, S.; Bazzi, M.; Bragadireanu, M.; Cargnelli, M.; Clozza, A.; Curceanu, C.; De Paolis, L.; Egger, J.-P.; et al. On the Importance of Electron Diffusion in a Bulk-Matter Test of the Pauli Exclusion Principle. *Entropy* **2018**, *20*, 515. [CrossRef]

Article

X-ray Detectors for Kaonic Atoms Research at DAΦNE

Catalina Curceanu [1], Aidin Amirkhani [2], Ata Baniahmad [2], Massimiliano Bazzi [1], Giovanni Bellotti [2], Carolina Berucci [1,†], Damir Bosnar [3], Mario Bragadireanu [4], Michael Cargnelli [5], Alberto Clozza [1], Raffaele Del Grande [1], Carlo Fiorini [2], Francesco Ghio [6], Carlo Guaraldo [1], Mihail Iliescu [1], Masaiko Iwasaki [7], Paolo Levi Sandri [1], Johann Marton [5], Marco Miliucci [1], Pavel Moskal [8], Szymon Niedźwiecki [8], Shinji Okada [7], Dorel Pietreanu [1,4], Kristian Piscicchia [1,9], Alessandro Scordo [1], Hexi Shi [1,‡], Michal Silarski [8], Diana Sirghi [1,4,*], Florin Sirghi [1,4], Magdalena Skurzok [8], Antonio Spallone [1], Hideyuki Tatsuno [10], Oton Vazquez Doce [1,11], Eberhard Widmann [5] and Johann Zmeskal [5]

[1] INFN, Laboratori Nazionali di Frascati, Frascati, 00044 Roma, Italy; catalina.curceanu@lnf.infn.it (C.C.); massimiliano.bazzi@lnf.infn.it (M.B.); carolinaberucci@gmail.com (C.B.); alberto.clozza@lnf.infn.it (A.C.); raffaele.delgrande@lnf.infn.it (R.D.G.); carlo.guaraldo@lnf.infn.it (C.G.); mihai.iliescu@lnf.infn.it (M.I.); paolo.levisandri@lnf.infn.it (P.L.S.); marco.miliucci@lnf.infn.it (M.M.); dorel.pietreanu@nipne.ro (D.P.); kristian.piscicchia@gmail.com (K.P.); alessandro.scordo@lnf.infn.it (A.S.); hexishi@lnf.infn.it (H.S.); fsirghi@lnf.infn.it (F.S.); antonio.spallone@lnf.infn.it (A.S.); oton.vazquez@universe-cluster.de (O.V.D.)

[2] Politecnico di Milano, Dipartimento di Elettronica, Informazione e Bioingegneria and INFN Sezione di Milano, 20133 Milano, Italy; aidin.amirkhani@polimi.it (A.A.); ata.baniahmad@polimi.it (A.B.); giovanni.bellotti@polimi.it (G.B.); carlo.fiorini@polimi.it (C.F.)

[3] Department of Physics, Faculty of Science, University of Zagreb, 10000 Zagreb, Croatia; bosnar@phy.hr

[4] Horia Hulubei National Institute of Physics and Nuclear Engineering, IFIN-HH, 077125 Magurele, Romania; mario.bragadireanu@nipne.ro

[5] Stefan-Meyer-Institut für Subatomare Physik, Vienna 1090, Austria; michael.cargnelli@oeaw.ac.at (M.C.); johann.marton@oeaw.ac.at (J.M.); eberhard.widmann@oeaw.ac.at (E.W.); johann.zmeskal@oeaw.ac.at (J.Z.)

[6] INFN Sez. di Roma I and Inst. Superiore di Sanita, 00161 Roma, Italy; francesco.ghio@lnf.infn.it

[7] RIKEN, Tokyo 351-0198, Japan; masa@riken.jp (M.I.); sokada@riken.jp (S.O.)

[8] The M. Smoluchowski Institute of Physics, Jagiellonian University, 30-348 Kraków, Poland; ufmoskal@googlemail.com (P.M.); szymonniedzwiecki@googlemail.com (S.N.); michal.silarski@uj.edu.pl (M.S.); mskurzok@gmail.com (M.S.)

[9] Museo Storico della Fisica e Centro Studi e Ricerche "Enrico Fermi", 00184 Roma, Italy

[10] Lund Univeristy, Faculty of Science, 22100 Lund, Sweden; hideyuki.tatsuno@gmail.com

[11] Excellence Cluster Universe, Technische Universiät München, 85748 Garching, Germany

* Correspondence: sirghi@lnf.infn.it

† Current address: Physics Department, University of Rome Tor Vergata, 00133 Rome, Italy.

‡ Current address: Institute of High Energy Physics, HEPHY, Vienna 1050, Austria.

Received: 19 February 2019; Accepted: 21 April 2019; Published: 25 April 2019

Abstract: This article presents the kaonic atom studies performed at the INFN National Laboratory of Frascati (Laboratori Nazionali di Frascati dell'INFN, LNF-INFN) since the opening of this field of research at the DAΦNE collider in early 2000. Significant achievements have been obtained by the DAΦNE Exotic Atom Research (DEAR) and Silicon Drift Detector for Hadronic Atom Research by Timing Applications (SIDDHARTA) experiments on kaonic hydrogen, which have required the development of novel X-ray detectors. The 2019 installation of the new SIDDHARTA-2 experiment to measure kaonic deuterium for the first time has been made possible by further technological advances in X-ray detection.

Keywords: kaonic atoms; strong interaction; X-ray detectors

---

1. Introduction

An exotic atom is an atomic system where an electron is replaced by a negatively charged particle, which could be a muon, a pion, a kaon, an antiproton, or a sigma hyperon, bound into an atomic orbit by its electromagnetic interaction with the nucleus.

Among exotic atoms, the hadronic ones, in which the electron is replaced by a hadron, play a unique role, since their study allows for the experimental investigation of the strong interaction described by Quantum Chromo Dynamics (QCD). The interaction is measured at threshold since the relative energy between the hadron forming the exotic atom and the nucleus is so small that it can be, for any practical purpose, neglected.

Experiments measuring kaonic atoms, in particular kaonic hydrogen, performed from the 70s through the 80s [1–3] have left the scientific community with a huge problem known as the "kaonic hydrogen puzzle": the measured strong interaction shift of the fundamental level with respect to the electromagnetic calculated value was positive, the resulting level more bound, which meant an attractive-type strong interaction between the kaon and the proton. This was in striking contradiction with the results of the analyses of low-energy scattering data, which found a repulsive-type strong interaction.

In this paper, we describe the DAΦNE Exotic Atom Research (DEAR) [4] and Silicon Drift Detector for Hadronic Atom Research by Timing Applications (SIDDHARTA) experiments [5] on kaonic hydrogen at the DAΦNE collider at the INFN National Laboratory of Frascati (Laboratori Nazionali di Frascati dell'INFN, LNF-INFN). These experiments have characterized the progress in detector development achieved in performing precision measurements of kaonic atoms.

DEAR has contributed to solve the "kaonic hydrogen puzzle", after the measurement of the KpX experiment at KEK [6], disentangling the full pattern of the *K*-series lines of kaonic hydrogen by employing charged-coupled devices (CCDs) that take advantage of their pixelized structure to obtain a powerful background reduction based on topological and statistical considerations.

SIDDHARTA used large area silicon drift detectors (SDDs) with microsecond timing capabilities. The main feature of the SDDs is the small value of the anode capacitance, enabling good resolution in energy and time. SIDDHARTA has performed the most precise measurement in the literature on kaonic hydrogen transitions.

In 2019, a new experiment, the SIDDHARTA-2 experiment, will be installed on DAΦNE to perform the first measurement of kaonic deuterium. The experimental challenge of the kaonic deuterium measurement is the yield, one order of magnitude less than kaonic hydrogen, and the even larger width. In order to satisfy the stringent requirements of the measurement, new monolithic SDD arrays have been developed with an improved technology, which increases the stability, optimizes the geometrical X-ray detection efficiency, and reduces the drift time.

Section 2 describes the kaonic atom measurements performed at DAΦNE by the DEAR experiment employing CCDs. Section 3 describes the measurements performed by the SIDDHARTA experiment employing SDDs. Section 4 looks at the future measurement of kaonic deuterium by the SIDDHARTA-2 experiment at DAΦNE. Conclusions are drawn in Section 5.

2. Kaonic Atom Measurements at DAΦNE Employing CCDs

2.1. Charge-Coupled Devices (CCDs)

Charge-coupled device (CCD) arrays are ideal detectors for a variety of X-ray imaging and spectroscopy applications, and in particular, in exotic atom research [7–9]. The CCD is essentially a silicon integrated circuit of the MOS type. The device consists of an oxide-covered silicon substrate with an array of closely spaced electrodes on top. Each electrode is equivalent to the gate of an MOS transistor. Signal information is carried in the form of electrons. The charge is localized beneath the electrodes with the highest applied potentials because the positive potential of an electrode causes the underlying silicon to be depleted to a certain depth and thus have a positive potential, which attracts the electrons. It is therefore common to say that the electrons are being stored in a "potential well". "Charge coupling" is a technique to transfer a signal charge from under one electrode to the next (Figure 1). This is achieved by also taking the voltage of the second electrode to a high level, then reducing the voltage of the first electrode. Therefore, by sequentially pulsing the voltages on the electrodes between high and low levels, charges can be made to pass down an array of many electrodes with hardly any loss and very little noise.

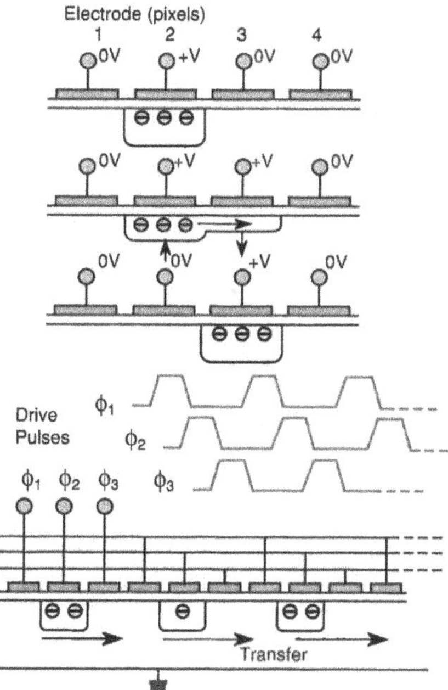

Figure 1. Charge signal transfer from one pixel to the next. One pixel contains three electrodes. The charges are in electrode No. 2, which has a voltage +V of 10 V. The voltage of electrode No. 3 is set to the same level as that of No. 2. Simultaneously, the voltage of No. 2 is reduced and the charges move to No. 3. A sequential pulsing of electrodes between two levels therefore allows a charge signal transfer over many electrodes (pixels).

CCDs are operated in vacuum and cooled down to approximately 160 K in order to limit dark current and therefore allow for up to several hours of exposure time. They operate in a similar way to conventional silicon solid state detectors in that the incoming X-rays, following absorption by photoelectric effect, are converted to electron–hole pairs where each pair requires 3.68 eV for its creation. In contrast to the visible photon case, the number of electrons created depends on the X-ray energy, and a good energy resolution can therefore be achieved.

The energy resolution of a CCD is given by:

$$\Delta E\ FWHM(eV) = 2.355 \times 3.68 (N^2 + \frac{FE}{3.68})^{1/2},$$ (1)

where N is the r.m.s. transfer and readout noise of the CCD, F is the Fano factor, and E is the X-ray energy. From the formula above, the best possible energy resolution with Si CCDs can be estimated by considering N^2 very small. The result is 70 eV FWHM at 2 keV and 140 eV FWHM at 8 keV. The characteristic parameters of CDDs are reported in Table 1.

Table 1. Comparison of X-ray detectors for kaonic atom research.

Detector	Si(Li)	CCD	SDD-JFET	SDD-CUBE
Effective area (mm^2)	200	724	3×100	8×64
Thickness (mm)	5	0.03	0.45	0.45
Energy resolution, FWHM, (eV) at 6 keV	410	150	160	140
Drift time (ns)	290	-	800	300
Experiment	KpX	DEAR	SIDDHARTA	SIDDHARTA-2
Reference	[6]	[4]	[5]	[10]

The identification of X-ray events and the determination of their energies is achieved by taking advantage of the pixel structure, which allows the application of a selection based on topological and statistical criteria [11]. This powerful background rejection tool is based on the fact that X-rays in the 1–10 keV energy range interact mainly via photoelectric effect and have a high probability of depositing all their energy in a single, or at most two, adjacent pixels, whereas the energy deposited from background particles (charged particles, gammas, neutrons) is distributed over several pixels (therefore called "cluster events"), which can be rejected (see Figure 2 (left)). A selected "single-pixel" (see Figure 2 (right)), a pixel with a charge content above a selected noise threshold, that is surrounded by eight neighbor pixels having a charge content below that threshold is considered to be an X-ray hit.

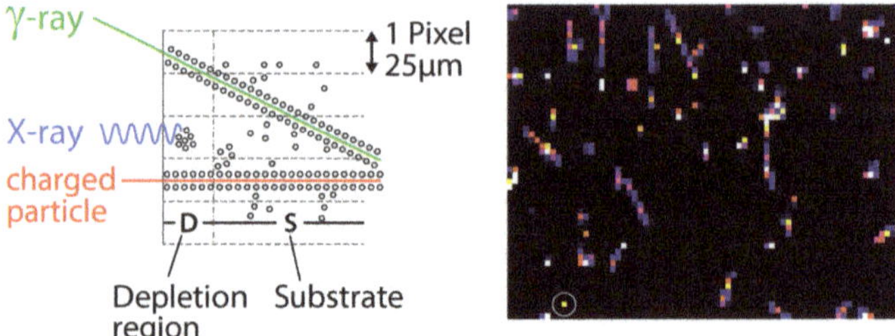

Figure 2. (**left**) Particle interactions and charge collection in a CCD detector; (**right**) example of an X-ray signal (inside the circle) in a CCD picture exposed during a data taking run.

2.2. The DEAR Kaonic Hydrogen Measurement at DAΦNE

After the KEK result, the primary goal of the DEAR experiment at the LNF-INFN e^+e^- DAΦNE collider was a precise determination of shift and broadening, due to strong interaction, of the fundamental level of kaonic hydrogen and a complete identification of the pattern of lines of the K-series transitions. The DEAR experiment took advantage of the clean (no contaminating particles in the beam), low-momentum (127 MeV/c), nearly monoenergetic ($\Delta p/p = 0.1\%$) beam of kaons from the decay of ϕ-mesons produced by e^+e^- collisions in the DAΦNE collider.

A shaped degrader of Kapton foils, from 150 µm up to 1200 µm thickness was used to put kaons at rest in the hydrogen atoms. The stopping efficiency was about 1%, with an intrinsic efficiency of almost 100%. At KEK, the 600 MeV kaon beam was produced by the 16 GeV proton beam of the KEK Proton Synchrotron on a thick target and then brought to rest in the hydrogen target using graphite degrader a few tens of cm thick. Due to the production mechanism, kaons were accompanied by pions in a ratio K/π equal to 1/90. The stopping efficiency was 0.06%, with an intrinsic efficiency of about 2% due to the broad energy distribution.

The DEAR setup consisted of three components: a kaon detector, a cryogenic target system, and an X-ray detection system. Figure 3 shows a schematic view of the setup. The whole setup was installed in one of the two interaction regions of DAΦNE.

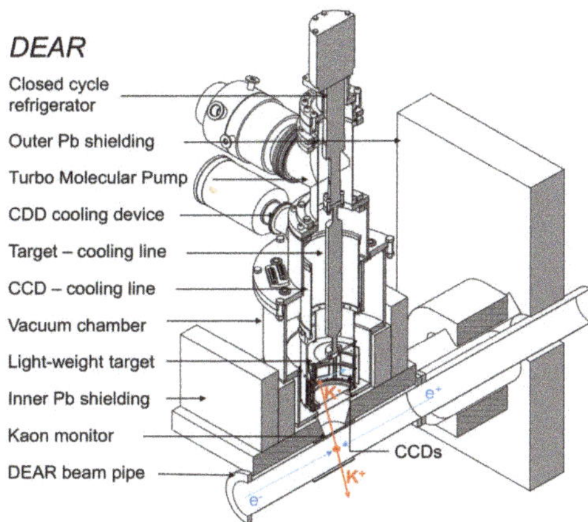

Figure 3. Schematic view of the DEAR experimental setup; only the right outer lead wall shielding is shown.

For X-ray detection, Marconi Applied Technologies CCD55-30 chips were selected. Each CCD55-30 chip has 1152×1242 pixels of $22.5 \times 22.5\ \mu m^2$, resulting in a total effective area of $7.24\ cm^2$ per chip. The depletion depth is about 30 μm. The study of transport and charge integration procedures has shown that the best results in terms of resolution and linearity can be obtained with a readout time of about 90 s. The evaluation of the occupancy effect indicates that a total exposure (readout plus static) of 120 s does not significantly reduce efficiency. Since the amount of data to be collected for each readout was relatively high, a period of 2 min was chosen. During the readout, the CCDs are exposed and since no imaging was necessary, the whole acquisition could be done in continuous readout (no static exposure). The target cell was surrounded by 16 CCDs covering a total area of $116\ cm^2$ and facing the cryogenic target cell. The CCD front-end electronics and controls and the data acquisition system were specially made for this experiment.

The number of hit pixels in a cluster categorizes the event type. In the DEAR analysis [11], events having 1 or 2 hit pixels were selected as X-ray events to increase both X-ray detection efficiency and the signal-to-noise ratio. The typical fraction of hit pixels per frame was about 3–5% so as to have an efficiency of hit recognition of about 98–99%. The X-ray detection efficiency as a function of energy and the X-ray event loss due to pile-up effect were calculated by means of Monte Carlo simulations and laboratory tests. The effect of applying charge-transfer efficiency corrections was an improvement in the resolution from 214 eV (FWHM) to 176 eV at the K_α line of Cu (8040 eV).

An energy calibration procedure based on fluorescence lines from setup materials excitation and the Ti and Zr foils was applied for each detector. Data from all individual detectors were then added. The overall resolution of the sum of detectors was determined: the values range from 130 eV (FWHM) for Ca K_α (3.6 keV) to 280 eV for Zr K_α (15.7 keV). The energy spectra consist of a continuous background component, fluorescence lines from setup materials, and kaonic hydrogen lines.

A measurement with non-colliding beams, i.e., e^+e^- beams separated in the interaction region, was performed. These data represented the so-called "no collisions background".

Two independent analyses were performed. The two analysis methods differ essentially in the background spectrum used. Analysis I used the bulk of no collisions data as the background spectrum. Analysis II used as the background spectrum the sum of kaonic nitrogen data [12], taken initially in order to optimize the kaon stopping distribution and to characterize the machine background, and a subset (low CCDs occupancy) of no collisions data. The two analyses gave consistent results. Figure 4 shows the kaonic hydrogen X-ray spectra for both analyses after continuous and structured background subtraction.

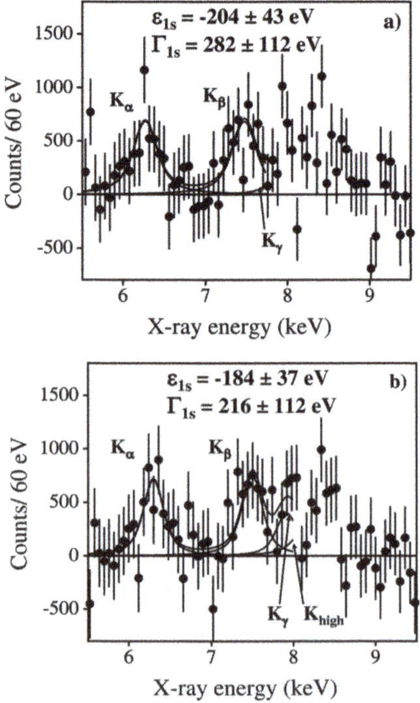

Figure 4. The DEAR kaonic hydrogen X-ray spectrum after continuous and structured background subtraction: (**a**) results of analysis I; (**b**) results of analysis II. The fitting curves of the various kaonic hydrogen lines are shown [4].

The resulting weighted average of the ground state shift ε_{1s} was

$$\varepsilon_{1s} = -193 \pm 37(stat) \pm 6(syst)\ eV. \tag{2}$$

The weighted 1s ground state width Γ_{1s} was

$$\Gamma_{1s} = 249 \pm 111(stat) \pm 30(syst)\ eV. \tag{3}$$

The DEAR results were consistent with the KEK measurement [6] to within 1σ of their respective errors. The repulsive-type character of the $K^- p$ strong interaction was confirmed.

The uncertainty of the DEAR results was about twice smaller than that of the KpX values. DEAR observed the full pattern of kaonic hydrogen K-lines, clearly identifying the K_α, K_β, and K_γ lines. The statistical significance of the summed intensities of the K-lines was 6.2 σ.

3. Kaonic Atom Measurements at DAΦNE Employing SDDs

3.1. Silicon Drift Detectors (SDDs)

The silicon drift detector, in its basic form proposed by Gatti and Rehak [13–15] in 1983, is a fully depleted detector in which an electric field parallel to the surface, created by properly biased contiguous field strips, drives signal charges towards a collecting anode (see Figure 5). The unique feature of this detector is the extremely low anode capacitance, which is moreover independent of the detector area. To take full advantage of the low output capacitance, the front-end n-channel JFET is integrated on the detector chip close to the n^+ implanted anode (Figure 5). They are located on the upper side of the device in the center of the p^+ field rings. Thus, stray capacitances of the various connections are minimized and a correct matching between detector and front-end electronics capacitance can be achieved.

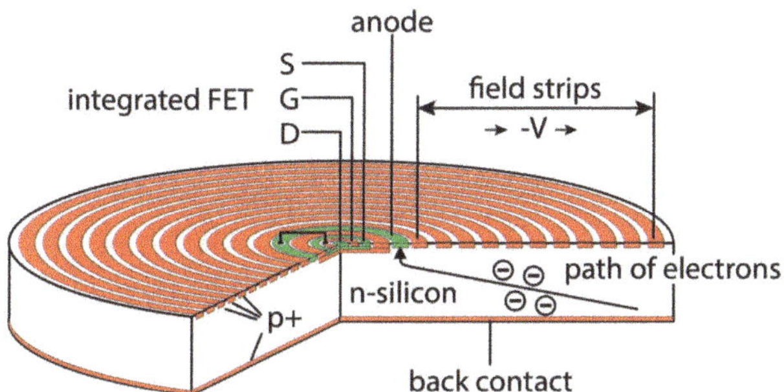

Figure 5. Cross section of a cylindrical silicon drift detector with integrated n-channel JFET. The gate of the transistor is connected to the collecting anode. The radiation entrance window for the ionizing radiation is the non-structured backside of the device.

The n-type device substrate is fully depleted by applying a negative voltage (with respect to the collecting anode) to the p^+ back contact and the p^+ field strips on the opposite side (see Figure 5). On this side, the negative bias of the p^+ rings progressively increases from the ring next to the anode to the farthest, outermost one. The maximum negative voltage of the outermost ring is about two times the voltage of the back contact. The minimum potential energy for electrons falls diagonally from the backside edge of the device to the readout electrode in the center of the upper side. Each electron generated inside the depleted detector volume by the absorption of ionizing radiation will therefore drift to the n^+ readout node. The generated holes are collected by the reverse-biased p^+ implanted regions.

As the device is fully depleted, the total thickness of 450 μm is sensitive to the absorption of ionizing radiation. For X-rays, this allows for more than 90% detection efficiency at 10 keV and more than 50% at 15 keV.

3.2. The SIDDHARTA Kaonic Hydrogen Measurement at DAΦNE

The SIDDHARTA experiment on DAΦNE at LNF-INFN [5] aimed to determine the kaonic hydrogen 1s shift and width with a higher precision than in DEAR [4], using large area SDDs.

Figure 6 shows a schematic view of the SIDDHARTA setup, which consisted of three main components: the kaon detector, X-ray detection system, and a cryogenic target system.

The SDDs in the SIDDHARTA experiment were developed within a European research project devoted to this experiment. Each of the 144 SDDs used in the apparatus had an area of 1 cm^2 and a thickness of 450 μm. Three cells were packed monolithically in one unit, as shown in Figure 7. The SDDs, operated at a temperature of ~170 K, had an energy resolution of 183 eV (FWHM) at 8 keV and a timing resolution below 1 μs, in contrast to the CCD detectors used in DEAR which had no timing capability. The characteristic parameters of the SDDs used by SIDDHARTA are reported in Table 1.

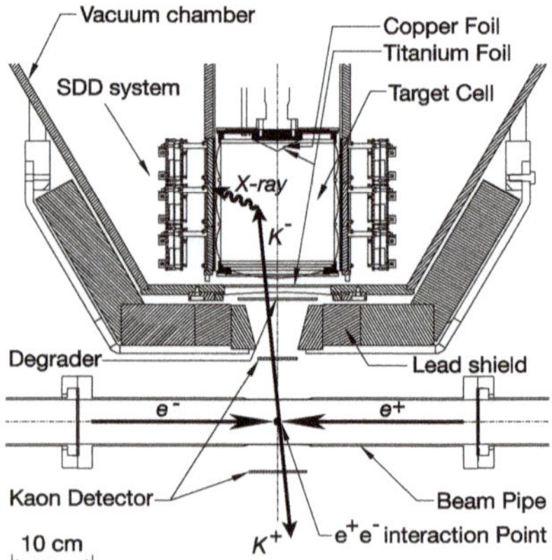

Figure 6. Schematic view of the SIDDHARTA setup [16].

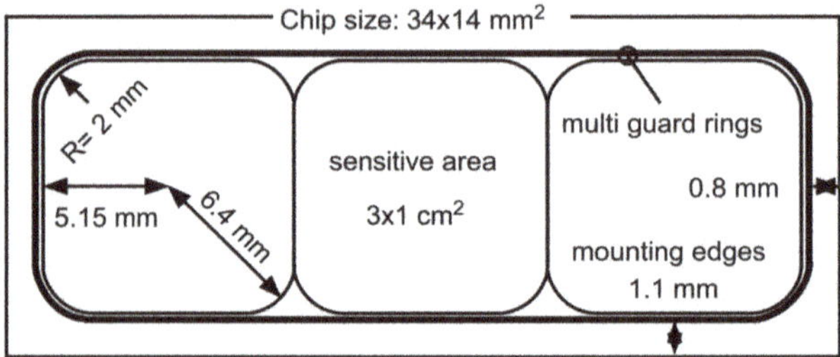

Figure 7. Schematic image of the SIDDHARTA SDDs. Each cell has an active area of 1 cm². Three cells are packed monolithically in one unit.

A trigger condition was used which took advantage of the characteristics of the back-to-back correlated charged kaon was production at DAΦNE, The time resolution of the SDDs allowed for the detection of the kaonic X-rays in coincidence with the back-to-back correlated $K^+ K^-$ pairs. There are two background sources in DAΦNE: backgrounds synchronous and asynchronous to the $K^+ K^-$ production. The main source is an asynchronous background and is due to electromagnetic showers originating from $e^+ e^-$ losses by the Touschek effect [17] and the interaction of the beams with the residual gas. The synchronous background, originating from particles produced by φ decay and secondary particles produced by kaon reactions as well as the decay particles of kaons, is small. Therefore, events related to charged-kaon production are selected only by demanding a triple coincidence of K^+, K^- and X-ray signal, so that the asynchronous background is rejected.

Figure 8 shows the time difference between the coincidence signals of the kaon monitor and the SDD events. The peak region contains the kaon-induced signals (kaonic atom X-rays) and background (gamma-rays and charged particles from the K^- interactions and K^+ decays). The tail of the distribution indicates the charge drift time in the SDDs. The time window indicated by the thick arrows was selected as synchronous events with charged kaons. The width of the timing window, from 2.4 µs to 4.6 µs, was adjusted to maximize the signal-to-background ratio and the statistical precision of the determined kaonic atom X-ray energy.

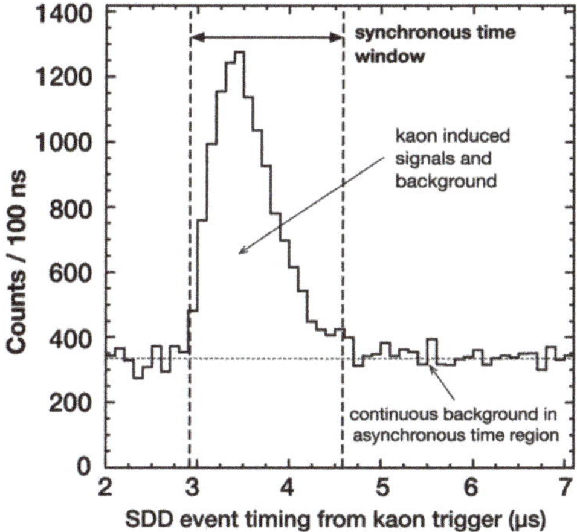

Figure 8. Timing spectrum of SDDs. The spectrum refers to a dataset with a target filled with helium-3. The peak corresponds to the time difference between the coincidence of the two scintillators of the kaon monitor and the X-ray events in the SDDs (triple coincidence). The peak region contains the kaon-induced signals and background. The time window indicated by arrows was selected to identify synchronous events with charged kaons. The continuous asynchronous background was reduced by this timing cut. From [16].

Using the coincidence between K^+K^- pairs and X-rays measured by SDDs, the main source of asynchronous background was drastically reduced, eventually resulting in an improvement of the signal-to-background ratio by more than a factor 10 with respect to the corresponding DEAR ratio of about 1/100.

The data acquisition system was built on a PCI bus base. The differential output signal from the readout chips was read out by ADC modules. Chip control, management of the memory and timing information, and the event construction were processed by FPGA modules. Energy data for all of the X-ray signals detected by the SDDs were recorded. In addition, a time difference between a trigger signal (generated by the coincidence signals in the kaon detector) and an X-ray signal in the SDDs was recorded using a clock with a frequency of 120 MHz whenever the coincidence signals occurred within a time window of 6 µs. This timing difference information is included inside the SDD data, which are used for the selection of kaon-timing events.

In the beginning of the SIDDHARTA runs, stability checks of SDD performance were examined using the kaonic helium X-ray lines by installing a thin Ti foil and a ^{55}Fe source inside the setup [18]. In Figure 9a, the peak position of the Mn K_α line (5.9 keV) as a function of time (about two weeks) is plotted, where the origin of the vertical axis is taken as an average of the Mn K_α peak positions in the whole dataset. A stability within ±2–3 eV was measured. This small instability was corrected to a fluctuation of ±0.5 eV in the data analysis, as indicated by "with correction" in the figure. In Figure 9b, the Mn K_α peak position against hit rates of the SDDs is plotted, where the origin of the vertical axis is taken as an average of the peak positions, and the horizontal axis is given by an arbitrary unit. The peak shift caused by hit rate dependency was found to be about ±2 eV, but this rate dependency was corrected from the relation between the rate and peak shift. With this correction, the rate dependency was corrected to be within ±0.5 eV. This stability is enough to determine the X-ray energy of the kaonic atom X-rays within the goal of the measurements.

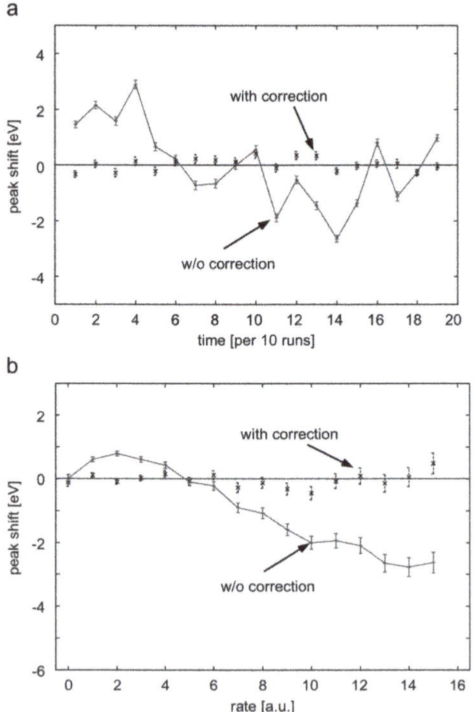

Figure 9. The X-ray peak shifts of the Mn K $_\alpha$ line (5.9 keV) as a function of time (**a**) and rate (**b**). The origin of the horizontal axis in the figures is the average of the data. With correction of the time dependency and rate dependency, a stability of ±0.5 eV was found [18].

In order to sum up the individual SDDs, the energy calibration of each single SDD was performed by periodic measurements of fluorescence X-ray lines from titanium and copper foils excited by an X-ray tube, with the e^+e^- beams in kaon production mode. A remote-controlled system moved the kaon detector out and the X-ray tube in once every 4 h for these calibration measurements. The refined in-situ calibration in gain (energy) and resolution (response shape) of the summed spectrum of all SDDs was obtained using titanium, copper, and gold fluorescence lines excited by the uncorrelated background without trigger and also using the kaonic carbon lines from wall stops in the triggered mode.

The use of the kaonic deuterium spectrum turned out to be essential to quantify the background lines originating from kaons captured in elements such as carbon, nitrogen, and oxygen contained in the setup materials, the deuterium data having no peak structures of K^-d X-rays due to their low yields and broad natural widths.

A global simultaneous fit of the hydrogen and deuterium spectra was performed. Figure 10a shows the residuals of the measured kaonic hydrogen X-ray spectrum after subtraction of the fitted background. *K*-series X-rays of kaonic hydrogen were clearly observed, while those for kaonic deuterium were not visible [5]. Figure 10b,c shows the fit result with the fluorescence lines from the setup materials and a continuous background. The vertical dot-dashed line in Figure 10 indicates the X-ray energy of kaonic hydrogen K_α calculated using the electromagnetic interaction only. When comparing the measured kaonic hydrogen K_α peak with the electromagnetic value, a repulsive-type shift (negative ε_{1s}) of the 1s energy level resulted.

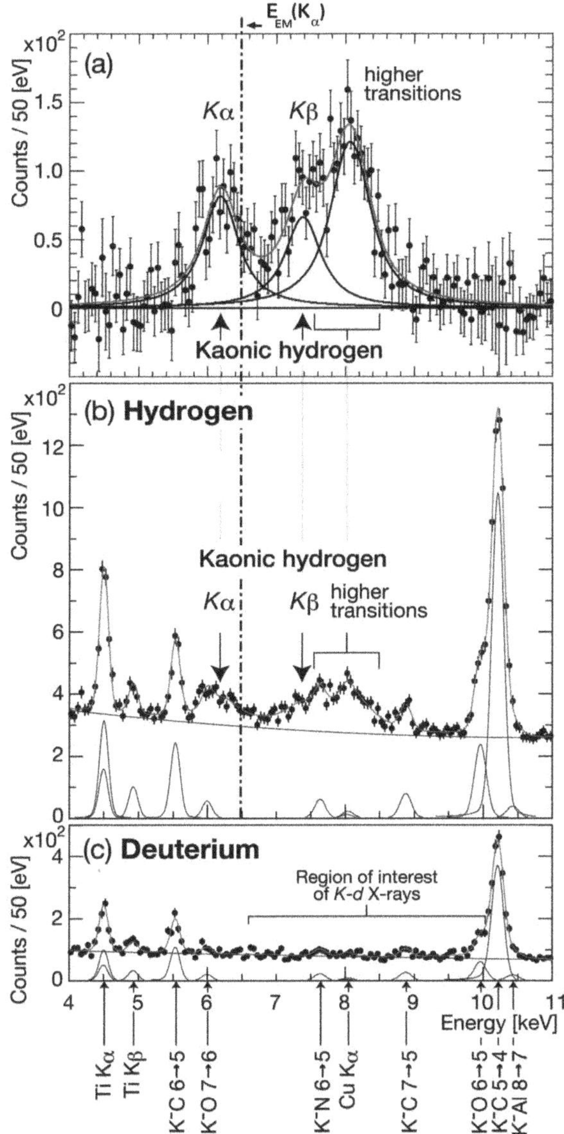

Figure 10. The global simultaneous fit of the X-ray energy spectra of hydrogen and deuterium data. (**a**) Residuals of the measured kaonic hydrogen X-ray spectrum after subtraction of the fitted background, clearly displaying the kaonic hydrogen K-series transitions. The fit components of the K^-p transitions are also shown, where the sum of the functions is drawn for the higher transitions (greater than K_β). (**b,c**) Measured energy spectra with fit lines. Fit components of the background X-ray lines and a continuous background are also shown. The dot-dashed vertical line indicates the e.m. value of the kaonic hydrogen K_α energy [5].

The 1s-level shift ε_{1s} and width Γ_{1s} of kaonic hydrogen were determined to be

$$\varepsilon_{1s} = -283 \pm 36(stat) \pm 6(syst) \, eV, \tag{4}$$

$$\Gamma_{1s} = 541 \pm 89(stat) \pm 22(syst) \; eV. \tag{5}$$

This is the most precise measurement of X-rays from the kaonic hydrogen atom performed so far.

4. Future Measurements on DAΦNE: Kaonic Deuterium

The kaonic deuterium X-ray measurement represents the most important experimental information missing in the low-energy antikaon–nucleus interactions field.

The experimental challenge of the kaonic deuterium measurement is the very small kaonic deuterium X-ray yield of one order of magnitude less than for hydrogen and the even larger width. There are two conditions which have to be fulfilled for the kaonic deuterium measurement at DAΦNE:

- A large area X-ray detector with good energy and timing resolution and stable working conditions. To meet the stringent requirements, new monolithic SDDs arrays have been developed. A difference with respect to the previously used SDDs is the change in the pre-amplifier system from a JFET structure on an SDD chip to a CMOS integrated charge-sensing amplifier (CUBE) [19]. For each SDD cell, this CUBE amplifier is placed on the ceramic carrier as close as possible to the anode of the SDD. The anode is electrically connected to the CUBE with a bonding wire. This makes the SDDs' performance almost independent of the applied bias voltages and increases their stability, even when exposed to high charged particle rates. A better drift time of 300 ns can be achieved with the newly developed SDDs compared to the previous ones ($\sim$800 ns) by changing the active cell area from 100 mm^2 to 64 mm^2 and by further cooling to 100 K. A new readout ASIC, named SFERA, has been developed to read out the SDDs of the SIDDHARTA-2 experiment [20]. The characteristic parameters of the SDDs used for SIDDHARTA-2 are reported in Table 1.
- Dedicated veto systems, to improve the signal-to-background ratio by at least one order of magnitude as compared to the kaonic hydrogen measurement performed by SIDDHARTA. Two special veto systems are foreseen for SIDDHARTA-2, consisting of an outer barrel of scintillator counters read by photomultipliers (PMs) and called Veto-1, and an inner ring of plastic scintillation tiles (SciTiles) read by silicon photomultipliers (SiPMs) placed as close as possible behind the SDDs for charged particle tracking, called Veto-2.

5. Conclusions

The experimental challenge in measuring kaonic atoms consists in the need to extract a weak signal under the high background conditions of the accelerators delivering kaon beams. This has required a continuous advance in X-ray detection that characterizes the era of precision measurements.

The first kaonic hydrogen X-ray measurement, which started the modern era of kaonic atoms research, made use of Si(Li) detectors in the KpX experiment at KEK (Japan). Charge-coupled devices (CCDs) were successfully used as X-ray detectors for the DEAR experiment at LNF-INFN. Finally, silicon drift detectors (SDDs) were developed for the SIDDHARTA program at DAΦNE. R&D work on SDDs continued, leading to an optimized detector for the future kaonic deuterium program at LNF-INFN. A comparison of the main characteristics of these detectors is given in Table 1.

Author Contributions: Conceptualization, C.C., C.G., J.M., M.I. (Mihail Iliescu), M.I. (Masaiko Iwasaki); software, C.B., M.B., M.C., R.D.G., M.I., M.M., S.O., D.P., K.P., H.S., D.S. and H.T.; methodology, A.A., A.B., M.B. (Massimiliano Bazzi), G.B., M.B. (Mario Bragadireanu), A.C., D.B., C.F., F.G., M.I., M.M., P.M., S.N., S.O., D.P., A.S., F.S., M.S. (Mical Silarski), M.S. (Magdalena Skurzok), H.T., O.V.D., E.W. and J.Z.; writing—original draft preparation, C.C. and C.G.; writing—review and editing, C.C. and D.S.

Funding: This research was funded by the Austrian Science Fund (FWF), P24756-N20; the Austrian Federal Ministry of Science and Research (BMBWK), 650962/0001 VI/2/2009; the Grantïn-Aid for Specially Promoted Research (20002003), MEXT, Japan; the Croatian Science Foundation under project IP-2018-01-8570; the Minstero degli Affari Esteri e della Cooperazione Internazionale, Direzione Generale per la Promozione del Sistema Paese (MAECI), Strange Matter Project; The Polish National Science Center, through grant No. UMO-2016/21/D/ST2/01155; and the Ministry of Science and Higher Education of Poland, grant no 7150/E-338/M/2018.

Acknowledgments: We thank C. Capoccia and G. Corradi from LNF-INFN, and H. Schneider, L. Stohwasser, and D. Pristauz-Telsnigg from the Stefan Meyer Institute for their fundamental contribution in designing and building the SIDDHARTA setup. We also thank the DAΦNE staff for the excellent working conditions and constant support.

Conflicts of Interest: The authors declare no conflict of interest.

References

1.   Davies, J.D.; Pyle, G.J.; Squier, G.T.A. Observation of kaonic hydrogen atom X-rays. *Phys. Lett. B* **1979**, *83*, 55. [CrossRef]

2.   Izycki, M.; Backenstoss, G.; Blum, L.T.P.; Guigas, R.; Hassler, N.; Koch, H.; Poth, H.; Fransson, K.; Nilsson, A.; Pavlopoulos, P.; et al. Results of the search for *K*-series X-rays from kaonic hydrogen. *Z. Phys. A* **1980**, *297*, 11. [CrossRef]

3.   Bird, P.M.; Clough, A.S.; Parke, K.; Pyle, G.; Squier, G.; Bair, S.; Batty, C.; Kilvington, A.; Russell, F.; Sharman, P. Kaonic Hydrogen atom X-rays. *Nucl. Phys. A* **1983**, *404*, 482. [CrossRef]

4.   Beer, G.; Bragadireanu, A.M.; Cargnelli, M.; Curceanu-Petrascu, C.; Egger, J.P.; Fuhrmann, H.; Guaraldo, C.; Iliescu, M.; Ishiwatari, T.; Itahashi, K.; et al. Measurement of the kaonic hydrogen X-ray spectrum. *Phys. Rev. Lett.* **2005**, *94*, 212302. [CrossRef] [PubMed]

5.   Bazzi, M.; Beer, G.; Bombelli, L.; Bragadireanu, A.; Cargnelli, M.; Corradi, G.; Curceanu, C.; d'Uffizi, A.; Fiorini, C.; Frizzi, T.; et al. A new measurement of kaonic hydrogen X-rays. *Phys. Lett. B* **2011**, *704*, 113. [CrossRef]

6.   Iwasaki, M.; Hayano, R.S.; Ito, T.M.; Nakamura, S.N.; Terada, T.P.; Gill, D.R.; Lee, L.; Olin, A.; Salomon, M.; Yen, S.; et al. Observation of kaonic hydrogen K_α X-rays. *Phys. Rev. Lett.* **1997**, *78*, 3067. [CrossRef]

7.   Varidel, D.; Bourquin, J.; Bovet, D.; Fiorucci, G.; Schenker, D. CCDs as low-energy X-ray detectors: II. Technical spects. *Nucl. Instrum. Meth. A* **1990**, *292*, 147. [CrossRef]

8.   Fiorucci, G.; Bourquin, J.P.; Bovet, D.; Bovet, E.; Egger, J.P.; Heche, C.; Nussbaum, C.; Schenker, D.; Varidel, D.; Vuilleumier, J.M. CCDs as low-energy X-ray detectors: I. General description. *Nucl. Instrum. Meth. A* **1990**, *292*, 141. [CrossRef]

9.   Egger, J.P.; Chatellard, D.; Jeannet, E. Progress in soft X-ray detection: The case of exotic hydrogen. *Part. World* **1993**, *3*, 139.

10.   Quaglia, R.; Bombelli, L.; Busca, P.; Fiorini, C.; Occhipinti, M.; Giacomini, G.; Ficorella, F.; Picciotto, A.; Piemonte, C. Silicon Drift Detectors and CUBE Preamplifiers for High-Resolution X-ray Spectroscopy. *IEEE Trans. Nucl. Sci.* **2015**, *62*, 221. [CrossRef]

11.   Ishiwatari, T.; Beer, G.; Bragadireanu, A.; Cargnelli, M.; Curceanu, C.; Egger, J.; Fuhrmann, H.; Guaraldo, C.; Iliescu, M.; Itahashi, K.; et al. New analysis method for CCD X-ray data. *Nucl. Instrum. Methods Phys. Res. A* **2006**, *556*, 509. [CrossRef]

12.   Ishiwatari, T.; Beer, G.; Bragadireanu, A.; Cargnelli, M.; Curceanu, C.; Egger, J.; Fuhrmann, H.; Guaraldo, C.; Iliescu, M.; Itahashi, K.; et al. Kaonic nitrogen X-ray transition yields in a gaseous target. *Phys. Lett. B* **2004**, *593*, 48. [CrossRef]

13.   Gatti, E.; Rehak, P. Semiconductor drift chamber—An application of a novel charge transport scheme. *Nucl. Instrum. Meth.* **1984**, *225*, 608. [CrossRef]

14.   Lechner, P.; Eckbauer, S.; Hartmann, R.; Krisch, S.; Hauff, D.; Richter, R.; Soltau, H.; Struder, L.; Fiorini, C.; Gatti, E.; et al. Silicon drift detectors for high resolution room temperature X-ray spectroscopy. *Nucl. Instrum. Meth. A* **1996**, *377*, 346. [CrossRef]

15.   Gatti, E.; Rehak, P. Review of semiconductor drift detectors. *Nucl. Instrum. Meth. A* **2005**, *541*, 47. [CrossRef]

16.   Bazzi, M.; Beer, G.; Bombelli, L.; Bragadireanu, A.; Cargnelli, M.; Corradi, G.; Curceanu, C.; d'Uffizi, A.; Fiorini, C.; Frizzi, T.; et al. First measurement of kaonic helium-3 X-rays. *J. Phys. Lett. B* **2011**, *697*, 199. [CrossRef] [PubMed]

17.   Bernardini, C.; Corazza, G.F.; Giugno, G.D.; Ghigo, G.; Haissinski, J.; Marin, P.; Querzoli, R.; Touschek, B. Lifetime and beam size in a storage ring. *Phys. Rev. Lett.* **1963**, *10*, 407. [CrossRef]

18.   Bazzi, M.; Beer, G.; Bombelli, L.; Bragadireanu, A.; Cargnelli, M.; Corradi, G.; Curceanu, C.; d'Uffizi, A.; Fiorini, C.; et al. Performance of silicon-drift detectors in kaonic atom X-ray measurements. *Nucl. Instrum. Methods Phys. Res. A* **2011**, *628*, 264. [CrossRef]

19.   Bombelli, L.; Fiorini, C.; Frizzi, T.; Alberti, R.; Longoni, A. "CUBE", A low-noise CMOS preamplifier as alternative to JFET front-end for high-count rate spectroscopy. In Proceedings of the 2011 IEEE Nuclear Science Symposium Conference Record, Valencia, Spain, 23–29 October 2011; p. 1972.
20.   Schembari, F.; Quaglia, R.; Bellotti, G.; Fiorini, C. SFERA: An integrated circuit for the readout of X and X-ray detectors. *IEEE Trans. Nucl. Sci.* **2016**, *63*, 1797–1807. [CrossRef]

Article

Energy Response of Silicon Drift Detectors for Kaonic Atom Precision Measurements

Marco Miliucci [1,2,*], **Mihail Iliescu** [1], **Aidin Amirkhani** [3], **Massimiliano Bazzi** [1], **Catalina Curceanu** [1], **Carlo Fiorini** [3], **Alessandro Scordo** [1], **Florin Sirghi** [1,4] **and Johann Zmeskal** [5]

[1]   Istituto Nazionale di Fisica Nucleare–Laboratori Nazionali di Frascati (LNF-INFN), 00044 Frascati RM, Italy; mihai.iliescu@lnf.infn.it (M.I.); Massimiliano.Bazzi@lnf.infn.it (M.B.); Catalina.Curceanu@lnf.infn.it (C.C.); Alessandro.Scordo@lnf.infn.it (A.S.); Sirghi.FlorinCatalin@lnf.infn.it (F.S.)

[2]   Department of Physics, Faculty of Science MM.FF.NN., University of Rome 2 (Tor Vergata), 00133 Rome, Italy

[3]   Politecnico di Milano, Dipartimento di Elettronica, Informazione e Bioingegneria and INFN Sezione di Milano, 20133 Milano, Italy; aidin.amirkhani@polimi.it (A.A.); carlo.fiorini@polimi.it (C.F.)

[4]   Horia Hulubei National Institute of Physics and Nuclear Engineering (IFIN-HH), 77125 Magurele, Romania

[5]   Stefan-Meyer-Institut für Subatomare Physik, 1090 Vienna, Austria; Johann.Zmeskal@oeaw.ac.at

*   Correspondence: Marco.Miliucci@lnf.infn.it

Received: 29 January 2019; Accepted: 6 March 2019; Published: 11 March 2019

Abstract: Novel, large-area silicon drift detectors (SDDs) have been developed to perform precision measurements of kaonic atom X-ray spectroscopy, for the study the $\overline{K}N$ strong interaction in the low-energy regime. These devices have special geometries, field configurations and read-out electronics, resulting in excellent performances in terms of linearity, stability and energy resolution. In this work the SDDs energy response in the energy region between 4000 eV and 12,000 eV is reported, revealing a stable linear response within 1 eV and good energy resolution.

Keywords: solid-state detectors; radiation detectors; photodetectors

---

1. Introduction

The main advantage of semiconductor X-ray detectors is the much lower energy required to create electron-hole pairs with respect to a gas detector, giving a greater number of charge carriers produced and, consequently, a better energy resolution. A silicon drift detector (SDD) consists of a double sided fully depleted silicon wafer with a cylindrical shape [1–3] where the n^- bulk is sided by a p^+ concentric ring strips and p^+ non-structured layer which forms the radiation entrance window. The radial drift field focuses the electrons produced by the absorbed radiation to the n^+ small anode placed in the centre of the p^+ strips side. The small value of the anode capacitance increases the amplitude of the output signal, giving good energy resolution and low noise in the subsequent electronic components also in high-count rate measurements. Since the anode capacitance is independent from the active area [4], these detectors can be built with a large active area. Furthermore, thanks to their reduced thickness, they can handle background events caused by high-energy particles still maintaining almost 100% efficiency for 8 keV X-rays.

The development of new SDD technologies dedicated to kaonic atom spectroscopy brought improvement in device performance with respect to past silicon detectors, allowing more precise and challenging measurements [5–11]. New monolithic SDD arrays have been developed by Fondazione Bruno Kessler (FBK, Italy), together with Politecnico di Milano (PoliMi, Italy), Istituto Nazionale di Fisica Nucleare-Laboratori Nazionali di Frascati (LNF-INFN, Italy) and Stefan Meyer Institute (SMI, Austria), to perform precise measurements of kaonic atom transitions at LNF-INFN and J-PARC.

The characterization and optimization of the energy response of the SDD detectors under temperature and voltage variations will be reported.

2. Materials and Methods

The SDD monolithic array is 450 μm thick and consists of a 2 × 4 matrix of square cells, each with an active area of 0.64 mm^2, glued on an Alumina carrier. The ceramic carrier provides a common polarization for all the devices to the following electrodes: The ring closest to the anode (R1), the outermost ring (RN) on the p$^+$ rings side and to the non-structured contact on the opposite side (Back). Keeping the voltage supplied to the R1 fixed at 15 V and moving the biasing of RN and Back (with fixed ratio RN/Back = 2), one can adjust the drift field inside the detector (V$_{Drift}$), optimizing the electron collection to the anode.

The ceramic carrier is screwed on top of an aluminium holder block, to protect the bonding and to cool the SDDs down to 120 K. The detector is placed inside a vacuum chamber (Figure 1a) with a Φ = 60 mm mylar window on top, a CONFLAT DN40 flange on one side for vacuum pumping and two Fischer (20 pin) vacuum connectors both for detector feed-throughs and temperature-pressure readings. The high vacuum (10^{-7} mbar) inside the chamber is obtained by a 80 L/s dry turbo molecular pump and a dry multistage "scroll" pump, which avoids detector surface contamination during the cooling. A dry air (high purity N$_2$ gas) flux keeps the system clean when the pumping is off and speeds up the detector heating from 250 K to room temperature without adding contaminants.

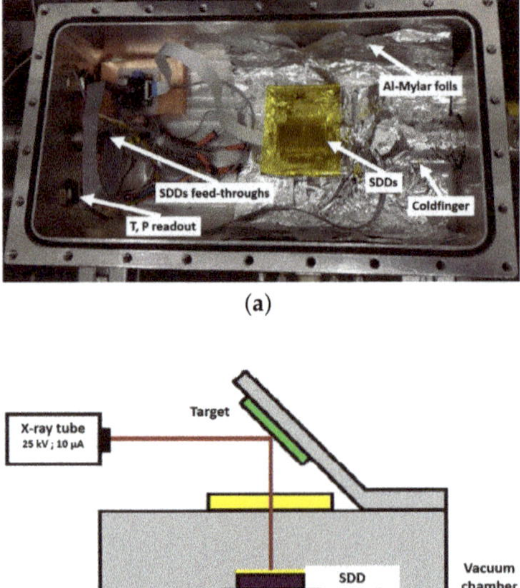

Figure 1. (a) Top view of SDDs array vacuum chamber; (b) Experimental setup in "reflection like" configuration.

The detector is screwed on a cold finger attached to an external cryostat, providing any intermediate temperature between 100 K and 250 K with a stability of 0.1 K. An additional detector surface protection, made by a suitable aluminium support, surrounds the detector and holds a 7.25 μm Mylar foil to prevent any residual gas condensation on its surface during the cooling.

Lastly, Al-Mylar foils shield the internal walls close to the cold finger from radiative heating.

The target is made of strips of titanium (Ti), iron (Fe) and copper (Cu) fixed on an epoxy plate encapsulating powder of potassium bromine (KBr) and provides fluorescent emission lines in the range 4500 eV to 12000 eV. The target is anchored on an aluminium support at 45° with respect to the detector surface, as showed in Figure 1b. A tungsten (W) anode X-ray tube (HXR 55-50-01, Oxford Instruments,

Abingdon, UK) shines on the target parallel to the entrance window, inducing the X-ray fluorescence. This "reflection-like" configuration optimizes the solid angle for both the detector placement and the X-ray activation beam, maximizing the signal over background ratio.

3. Results and Discussion

3.1. Energy Calibration

The collected spectra have been calibrated in energy using the K_α peaks of the excited elements of the target, fitted with a single function consisting in a sum of Gaussians and tails for the fluorescence lines and an exponential function to describe the background [12]. As an example, Figure 2 shows the fitted spectrum used for the calibration of a fluorescence spectrum obtained at $T_{SDD} = 121.6 \pm 0.1$ K and $V_{Drift} = 140.0 \pm 0.1$ V. A linear fit interpolates the calibration points, whose coordinates are the theoretical (P_i^{label}) and the experimental values of each K_α peak, as showed in Figure 3a. The slope of the function, in eV/ch units, gives the gain parameter (g) of the spectrum. The difference between each K_α calibrated position (P_i^{cal}), with respect to its corresponding theoretical value, gives the residual plot shown in the Figure 3b, used for the evaluation of the system linearity. The distribution reveals that the distance of each calibrated peak from the theoretical value is below 1 eV, so the systematic error in the evaluation of the peak position over the whole energy region is better than 1 eV.

The procedure has been performed for a set of measurements, collected by varying V_{Drift}, evaluating the linearity parameter (l) by using Equation (1):

$$l = \frac{\sqrt{\sum_1^4 (P_i^{label} - P_i^{cal})^2}}{4} \tag{1}$$

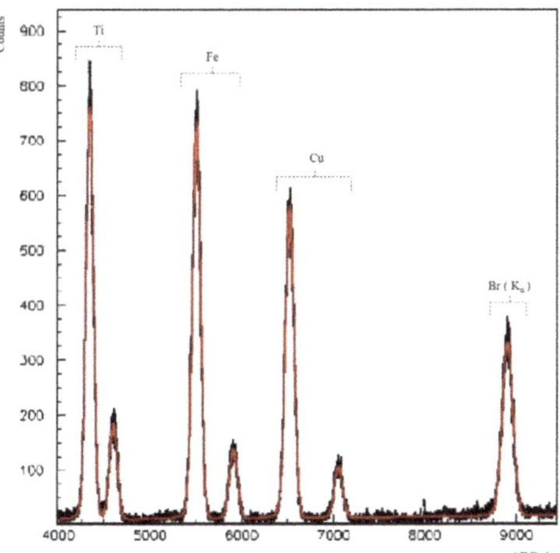

Figure 2. Fit of the fluorescence spectrum collected with $T_{SDD} = 121.6 \pm 0.1$ K and $V_{Drift} = 140.0 \pm 0.1$ V.

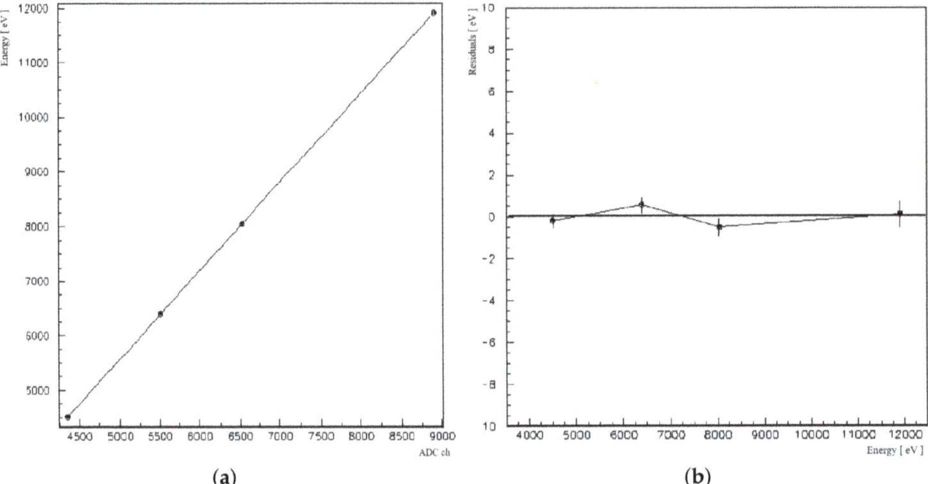

Figure 3. (**a**) Linear calibration of the spectrum using K$_\alpha$ peaks of the four elements; (**b**) Residuals plot.

Keeping fixed the R1 polarization at 15.0 ± 0.1 V and the SDD temperature at 121.6 ± 0.1 K, a linearity and gain study has been performed in the V$_{Drift}$ range between 100–180 V, with the results summarized in Figure 4. Starting from the lower voltage up to 175.0 ± 0.1 V the distribution of the points is stable for both residuals and gain. For the points below 175 V, the mean value of the linearity is set around 1 eV, consistent with the previous evaluation. By moving the V$_{Drift}$ up to 180 V, the excellent linear response of the system is suddenly lost, resulting in an increment of the residuals up to 2.5 eV.

This trend is reflected also in the gain plot, which presents a sharp rise at V$_{Drift}$ = 180.0 ± 0.1 V. It reveals that the peak position on the ADC spectra are shifted downward with respect to the lower voltages, as shown also in Figure 5a, where the overlap of two non-calibrated spectra collected at different V$_{Drift}$ are presented. This behaviour indicates a reduced amount of charge collected to the anode. The incomplete charge collection is related to the high voltage applied, which moves the focus of the drift field away from the anode [13]. In addition to the previous considerations, Figure 5b shows that the spectrum collected 2 V above 180 V, presents also a sensible energy resolution worsening.

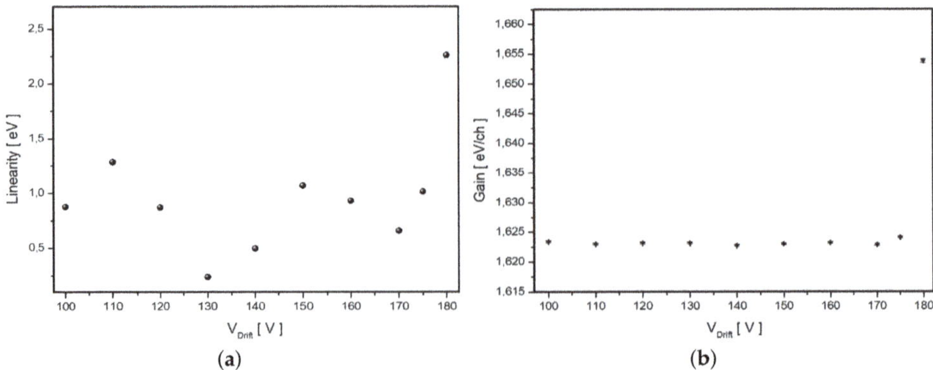

Figure 4. (**a**) Linearity parameter (*l*); (**b**) Gain of the system under voltage scan.

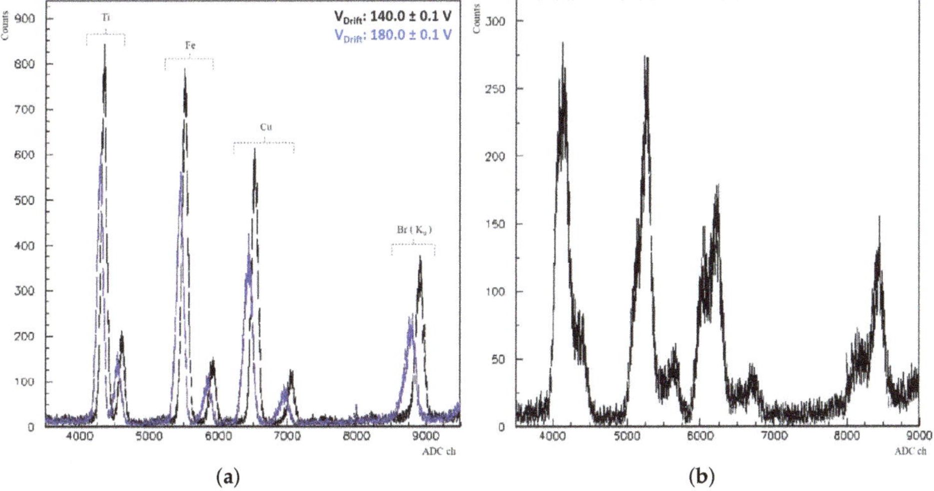

(**a**)   (**b**)

Figure 5. (**a**) The overlap of spectrum collected at V_{Drift} = 140.0 ± 0.1 V (black) and V_{Drift} = 180.0 ± 0.1 V (blue) shows a shift downward of the spectrum; (**b**) The energy resolution worsening for the spectrum collected at V_{Drift} = 182.0 ± 0.1 V reveals that the electron collection is not focused to the anode.

3.2. Stability

An important aspect of the SDD system to be used in kaonic atom experiments is its stability over a long running period. To evaluate it, a spectrum collected over two days of running has been sampled in intervals of two hours.

The following parameters of the data taking were stable during the test:

- X-ray tube voltage: 23.0 ± 0.1 kV;
- X-ray tube current: 10.0 ± 0.1 µA;
- SDDs temperature: 121.6 ± 0.1 K;
- SDDs V_{Drift}: 140.0 ± 0.1 V;

The stability has been verified from the calibrated values of the Fe K$_\alpha$ peak, using the K$_\alpha$ lines of titanium and bromine to obtain the ADC to eV conversion of the spectrum.

The results are plotted in Figure 6.

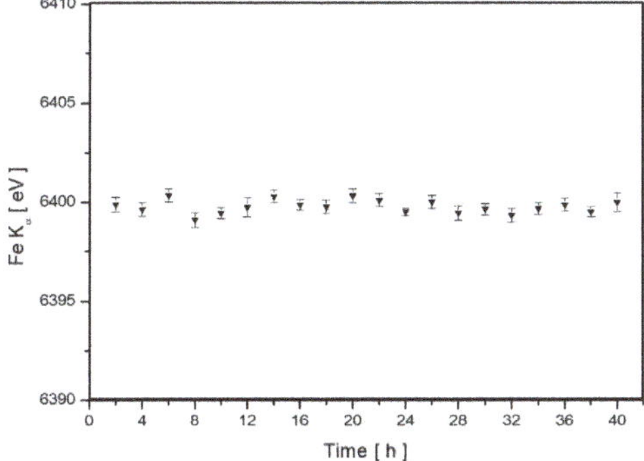

Figure 6. Calibrated position of Fe K$_\alpha$ peak during the stability test run.

The values obtained from the analysis are fluctuating around the reference value of the Fe K_α line (the weighted mean value of Fe $K_{\alpha 1}$ and $K_{\alpha 2}$ is 6399.64 eV). An oscillating trend, with an amplitude around 1 eV is present. This can be due to temperature variations between day and night in the laboratory, which affect the readout electronic. The mean value of the calibrated peak position over the whole range is 6399.8 $\pm$ 0.4 eV and differs by 0.2 eV from the reference value, consistently with the systematic error previously evaluated.

3.3. Energy Resolution

The energy resolution reflects the detector accuracy in the determination of the incoming radiation energy. For the SDDs, the total energy resolution (ΔE_{tot}^2) results from the sum of three distinct contributions, as described by Equation (2):

$$\Delta E_{tot}^2 = \Delta E_{intr}^2 + \Delta E_{e.n.}^2 + \Delta E_{c.c.}^2 \tag{2}$$

where:

- ΔE_{intr}^2 is the intrinsic spread given by the statistical fluctuation in the number of the charge created by the incoming radiation. For a Gaussian distribution, it is due to the energy of the line (E_i), the electron hole pair creation energy (ϵ) and the correction of the Fano factor (F), accordingly to the relation $\Delta E_{intr}^2 = (2.35)^2 \cdot F \cdot \epsilon \cdot E_i$;
- $\Delta E_{e.n.}^2$ is the thermal and electronic noise contribution;
- $\Delta E_{c.c.}^2$ is due to the incomplete charge collection;

As showed in the previous analysis concerning the linear response of the detectors, the contribution due to the incomplete charge collection is negligible below 180 V.

The energy resolution has been investigated as a function both of drift field and temperature, and, given ϵ [14], the Fano factor has been extracted from the Fe K_α and Ti K_α peaks.

Figure 7a shows the FWHM, expressed in eV, of Fe K_α and Ti K_α in the range between 100.0 V and 170.0 V, for a fixed temperature of 123.1 $\pm$ 0.1 K. The red dots are for Fe K_α FWHM, while the blue dots are for Ti K_α FWHM. The width of the peaks is stable over the scan, meaning that the energy resolution of the SDD is not sensitive to the bias of the electrodes inside a wide range of applied V_{Drift}. The intrinsic contribution of the detector resolution, resulting from the obtained values of the Fano factor, reflects that the Gaussian widening of the peak, due to the pair creation statistics, is not affected by the drift field in the investigated range. The mean value experimentally obtained for the Fano factor is 0.118 $\pm$ 0.009.

A temperature scan has been done fixing V_{Drift} at 140.0 $\pm$ 0.1 V and increasing the detector temperature from 123.1 $\pm$ 0.1 K up to 233.1 $\pm$ 0.1 K. Figure 8a shows that the energy resolution is quite constant below 170 K, while a relevant peak broadening is detected starting from 180 K up. For this study, the Fano factor results to be stable (Figure 8b), with a mean value of 0.116 $\pm$ 0.009, compatible to the previous result. Since the intrinsic contribution of the peak widening is not affected by the temperature as demonstrated by the Fano factor distribution, the energy resolution worsening with the temperature is associated to an higher thermal noise inside the detector. Its increment is due to the larger number of carriers created by thermal excitation (leakage current). Thus, a progressive cooling of the device ensures lower leakage current, as well as higher performances in terms of energy resolution.

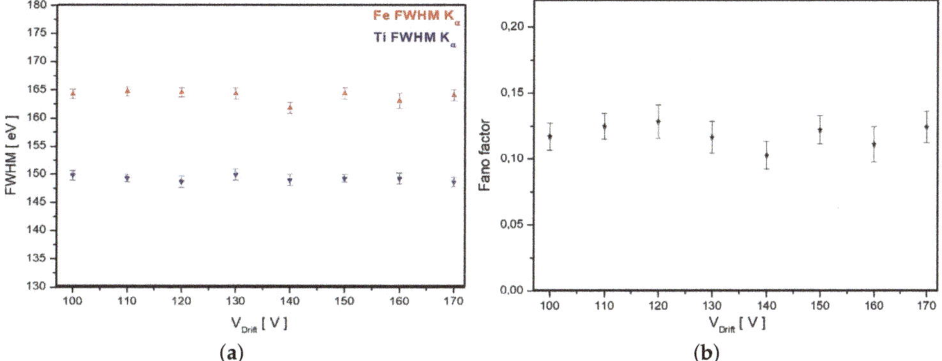

Figure 7. Energy response of the detector under the voltage scan. (**a**) Distribution of energy resolution of Ti K_α (blue) and Fe K_α (red) for different drift voltages; (**b**) Fano factor values at different drift voltages.

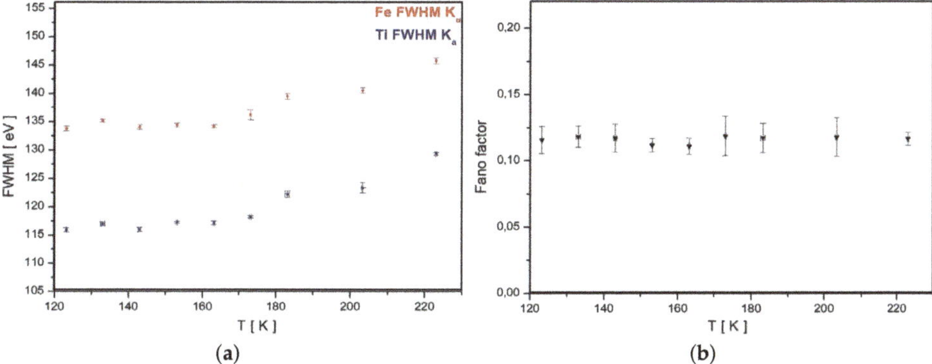

Figure 8. Energy response of the detector under the temperature scan. (**a**) The widening of Ti K_α (blue) and Fe K_α (red) peaks are related to the increasing of the detector temperature; (**b**) Plot of the detector Fano factor at different temperatures reveals that the intrinsic contribution is not affected by the temperature within the experimental error.

4. Conclusions

The work presented in this paper qualifies and optimizes the new technology of silicon drift detectors to be used for precision measurements of kaonic atom X-ray transitions. The detectors are stable and linear within 1 eV in the energy range between 4500–12000 eV, which means that the relative systematic error is at the level of 10^{-4}. The voltage scan grants that the common polarization of the eight units of the matrix has no drawbacks in terms of linearity for a wide range, below an upper limit set by a not efficient charge collection to anode. Likewise, the energy resolution is not affected by the variation of the drift field inside the detector, in the range of the stable linear response. The intrinsic resolution, extracted by the width of the Ti and Fe K_α peaks, results as independent from the collecting field, giving a mean value for the Fano factor equal to 0.118 ± 0.009.

A temperature scan shows that the widening of the peaks is due to an increment of the leakage current inside the detector, testifying that the cooling of the device ensures the best performances. The intrinsic contribution is constant, resulting in a Fano factor value of 0.116 ± 0.009.

The analysis of the silicon drift detector energy response presented in this work optimizes the working conditions of the device, indicating them as excellent candidates for precision measurements of kaonic atom X-ray transitions.

Author Contributions: Conceptualization, M.M. and M.I.; software, M.I.; validation, M.M., M.I. and C.C.; formal analysis, M.M. and A.S.; investigation, M.M.; resources, M.M., A.A., M.B., F.S., C.F. and J.Z.; data curation, M.M.; writing—original draft preparation, M.M.; writing—review and editing, M.M.; administration, M.M., M.I. and C.C.; funding acquisition, C.C., C.F. and J.Z.

Funding: Part of this work was supported by the Austrian Science Fund (FWF): [P24756-N20]; Austrian Federal Ministry of Science and Research BMBWK 650962/0001 VI/2/2009; the Croatian Science Foundation, project IP2018-01-8570; Ministero degli Affari Esteri e della Cooperazione Internazionale, Direzione Generale per la Promozione del Sistema Paese (MAECI), StrangeMatter project; Polish National Science Center through grant no. UMO-2016/21/D/ST2/01155; Ministry of Science and Higher Education of Poland grant no 7150/E338/M/2018.

Conflicts of Interest: The authors declare no conflict of interest.

References

1. Gatti, E.; Rehak, P. Semiconductor drift chamber—An application of a novel charge transport scheme. *Nucl. Instrum. Meth. Phys. Res.* **1984**, *225*, 608–614. [CrossRef]

2. Gatti, E.; Rehak, P. Silicon drift chambers—First results and optimum processing of signals. *Nucl. Instrum. Meth. Phys. Res. A* **1984**, *226*, 129–141. [CrossRef]

3. Lechner, P.; Eckbauer, S.; Hartmann, R.; Krisch, S.; Hauff, D.; Richter, R.; Soltau, H.; Strüder, L.; Fiorini, C.; Gatti, E.; et al. Silicon drift detectors for high resolution room temperature X-ray spectroscopy. *Nucl. Instrum. Meth. Phys. Res. A* **1996**, *377*, 346–351. [CrossRef]

4. Rehak, P.; Gatti, E.; Longoni, A.; Kemmer, J.; Holl, P.; Klanner, R.; Lutz, G.; Wylie, A. Semiconductor drift chambers for position and energy measurements. *Nucl. Instrum. Meth. Phys. Res. A* **1985**, *235*, 224–234. [CrossRef]

5. Iwasaki, M.; Hayano, R.S.; Ito, T.M.; Nakamura, S.N.; Terada, T.P.; Gill, D.R.; Lee, L.; Olin, A.; Salomon, M.; Yen, S.; et al. Observation of Kaonic Hydrogen K_α X-Rays. *Phys. Rev. Lett.* **1997**, *78*, 3067–3069. [CrossRef]

6. Bazzi, M.; Beer, G.; Bombelli, L.; Bragadireanu, A.M.; Cargnelli, M.; Corradi, G.; Curceanu, C.P.; d'Uffizi, A.; Fiorini, C.; Frizzi, T.; et al. A new measurement of kaonic hydrogen X-rays. *Phys. Lett. B* **2011**, *704*, 113–117. [CrossRef]

7. Okada, S.; Beer, G.; Bhang, H.; Cargnelli, M.; Chiba, J.; Choi, S.; Curceanu, C.; Fukuda, Y.; Hanaki, T.; Hayano, R.S.; et al. Precision measurement of the $3d{\rightarrow}2p$ X-ray energy in kaonic ^{4}He. *Phys. Lett. B* **2007**, *653*, 387–391. [CrossRef]

8. Bazzi, M.; Beer, G.; Bombelli, L.; Bragadireanu, A.M.; Cargnelli, M.; Corradi, G.; Curceanu, C.; d'Uffizi, A.; Fiorini, C.; Frizzi, T.; et al. Kaonic helium-4 X-ray measurement in SIDDHARTA. *Phys. Lett. B* **2009**, *681*, 310–314. [CrossRef]

9. Bazzi, M.; Beer, G.; Bombelli, L.; Bragadireanu, A.M.; Cargnelli, M.; Corradi, G.; Curceanu, C.; d'Uffizi, A.; Fiorini, C.; Frizzi, T.; et al. First measurement of kaonic helium-3 X-rays. *Phys. Lett. B* **2011**, *697*, 199–202. [CrossRef] [PubMed]

10. Bazzi, M.; Beer, G.; Bellotti, G.; Berucci, C.; Bragadireanu, A.M.; Bosnar, D.; Cargnelli, M.; Curceanu, C.; Butt, A.D.; d'Uffizi, A.; et al. K-series X-ray yield measurement of kaonic hydrogen atoms in a gaseous target. *Nucl. Phys. A* **2016**, *954*, 7–16. [CrossRef]

11. Bazzi, M.; Beer, G.; Berucci, C.; Bragadireanu, A.M.; Cargnelli, M.; Curceanu, C.; d'Uffizi, A.; Fiorini, C.; Ghio, F.; Guaraldo, C.; et al. L-series X-ray yields of kaonic ^{3}He and ^{4}He atoms in gaseous targets. *Eur. Phys. J. A* **2014**, *50*, 1–4. [CrossRef]

12. Van Gysel, M.; Lemberge, P.; Van Espen, P. Implementation of a spectrum fitting procedure using a robust peak model. *X-ray Spectrom.* **2003**, *32*, 434–441. [CrossRef]

13. Bertuccio, C.; Castoldi, A.; Longoni, A.; Sampietro, M.; Gauthier, C. New electrode geometry and potential distribution for soft X-ray drift detectors. *Nucl. Instrum. Meth. Phys. Res. A* **1992**, *312*, 613–616. [CrossRef]

14. Mazziotta, M.N. Electron–hole pair creation energy and Fano factor temperature dependence in silicon. *Nucl. Instrum. Meth. Phys. Res. A* **2008**, *584*, 436–439. [CrossRef]

Article

Pyrolitic Graphite Mosaic Crystal Thickness and Mosaicity Optimization for an Extended Source Von Hamos X-ray Spectrometer

Alessandro Scordo [1],*, Catalina Curceanu [1], Marco Miliucci [1], Florin Sirghi [1,2] and Johann Zmeskal [3]

[1]   Laboratori Nazionali di Frascati (INFN), Via E. Fermi 40, 00044 Frascati, Italy;
     catalina.curceanu@lnf.infn.it (C.C.); Marco.Miliucci@lnf.infn.it (M.M.); fsirghi@lnf.infn.it (F.S.)
[2]   Horia Hulubei National Institute of Physics and Nuclear Engineering (IFIN-HH), Strada Reactorului 30,
     MG-6 Măgurele, Romania
[3]   Stefan-Meyer-Institut für Subatomare Physik, Boltzmanngasse 3, 1090 Vienna, Austria;
     Johann.Zmeskal@oeaw.ac.at
*   Correspondence: alessandro.scordo@lnf.infn.it

Received: 28 January 2019; Accepted: 29 March 2019; Published: 3 April 2019

Abstract: Bragg spectroscopy, one of the best established experimental techniques for high energy resolution X-ray measurements, has always been limited to the measurement of photons produced from well collimated (tens of microns) or point-like sources; recently, the VOXES collaboration at INFN National Laboratories of Frascati developed a prototype of a high resolution and high precision X-ray spectrometer working also with extended isotropic sources. The realized spectrometer makes use of Highly Annealed Pyrolitic Graphite (HAPG) crystals in a "semi"-Von Hamos configuration, in which the position detector is rotated with respect to the standard Von Hamos one, to increase the dynamic energy range, and shows energy resolutions at the level of 0.1% for photon energies up to 10 keV and effective source sizes in the range 400–1200 µm in the dispersive plane. Such wide effective source dimensions are achieved using a double slit system to produce a virtual point-like source between the emitting target and the crystal. The spectrometer performances in terms of reflection efficiency and peak resolution depend on several parameters, among which a special role is played by the crystal mosaicity and thickness. In this work, we report the measurements of the $Cu(K_{\alpha1,2})$ and the $Fe(K_{\alpha1,2})$ lines performed with different mosaicity and thickness crystals in order to investigate the influence of the parameters on the peak resolution and on the reflection efficiency mentioned above.

Keywords: X- and γ-ray instruments; X- and γ-ray sources, mirrors, gratings, and detectors; X-ray and γ-ray spectrometers; optical materials; X-ray diffraction; optical instruments and equipment

PACS: 07.85.-m; 07.85.Fv; 07.85.Nc; 42.70.-a; 61.10.Nz; 07.60.-j

---

1. Introduction

High precision measurements of soft X-rays represent still today a very big challenge; nevertheless, such kind of measurements are strongly demanded in many fields of fundamental science, from particle and nuclear physics to quantum mechanics, as well as in astronomy and in several applications using synchrotron light sources or X-ray Free Electron Lasers (X-FEL) beams, in biology, medicine and industry. For several of these applications, in particular for nuclear physics experiments like those involving the measurement of exotic atoms' radiative transitions, the detection of photons isotropically emitted from extended (non point-like) targets is required. Among the typically used solid state large area spectroscopic detectors, the Silicon Drift Detectors (SDDs), recently employed by the SIDDHARTA experiment [1] for exotic atoms' transition lines measurements at the $DA\Phi NE\ e^+e^-$ collider of the INFN National Laboratories of Frascati [2] represent the best options for wide and isotropic targets, in terms of energy resolution. The intrinsic resolution of such kind of detectors is nevertheless limited to $\simeq$120 eV FWHM at 6–8 keV by the Fano Factor, making them unsuitable for those cases in which the photon energy has to be measured with a precision below 1 eV.

The superconducting microcalorimeters' Transition Edge Sensors (TES), recently developed at NIST [3], represent a real possibility to obtain few eV FWHM at 6 keV; however, these kinds of detectors still have some limitations: a very small active area, prohibitively high costs of the complex cryogenic system needed to reach the operational temperature of $\simeq$50 mK, and a response function which is still not properly under control.

A further alternative is represented by Bragg spectroscopy, one of the best established high resolution X-ray measurement techniques, where the X-ray energy spectrum can be otained exploiting photon reflection on crystals following the Bragg rule $n\lambda = 2d \sin\theta$, where n is an integer number (order of reflection), λ is the photon wavelength, d is the crystal lattice constant and θ is the photon impinging angle on the crystal surface. If a monochromatic parallel X-ray beam has to be measured, the spectrum is obtained by several measurements with different θ values around the nominal Bragg angle θ_B; on the contrary, if one needs to measure a polychromatic or isotropic beam of photons, the spectrum is obtained in one shot with a single θ using a position detector. In the first case, higher resolutions can be obtained with the drawback of a longer data taking; in the second case, in which the measurements presented in this work fall into, the faster exposure time is balanced by a higher background due to the beam divergence and the source size. As a consequence, when the photons emitted from extended isotropic sources (like a gaseous or liquid target) have to be measured, this method has been until now ruled out by the constraint to reduce the dimension of the target to a few tens of microns [4,5].

Experiments performed in the past at the Paul Scherrer Institute (PSI), measuring pionic atoms [6,7], pioneered the possibility to combine Charged Coupled Device detectors (CCDs) with silicon crystals, but the energy range achievable with that system was limited to few keV due to the crystal structure, and the silicon low intrinsic reflection efficiency required the construction of a very large spectrometer. The possibility to perform other fundamental measurements, like the precision determination of the K^- mass measuring the radiative kaonic nitrogen transitions at the $DA\Phi NE$ collider [8], has been also investigated, but the estimated efficiency of the proposed spectrometer was not sufficient to reach the required precision.

In the last several decades, the development of the Pyrolitic Graphite mosaic crystals [9–11] renewed the interest in Bragg spectrometers as possible candidates also for millimetric isotropic sources' X-ray measurements. Mosaic crystals consist of a large number of nearly perfect small pyrolitic graphite crystallites, randomly misoriented around the lattice main direction; the FWHM of this random angular distribution is called mosaicity (ω_{FWHM}) and it makes it possible that even a photon not reaching the crystal with the exact Bragg energy-angle relation can find a properly oriented crystallite and be reflected [12]. This, together with a lattice spacing constant of 3514 Å, enables them to be highly efficient in diffraction in the 2–20 keV energy range, for the $n = 1$ reflection order, while higher energies can be reached at higher reflection orders.

From the definition of mosaicity, it follows that those photons which find a properly oriented crystallite in the inner part of the crystal are reflected on the position detector slightly shifted with respect to the one coming from a surface crystallite; as a consequence of this unfocusing process, the peak resolution is worsened with respect to a standard non-mosaic crystal. For the same principle, for a given mosaicity, the thickness of the graphite crystal plays a fundamental role; for higher thickness, the reflection efficiency is enhanced at the expence of the peak resolution. This interplay between the above mentioned parameters has been investigated in the past with single or double crystal spectrometers providing parallel and monochromatic sources of few tens of microns. However, in spite of the results published and available in literature [12–16], it is still useful and interereresting to investigate how thickness and mosaicity influence the spectrometer response when almost millimetric sources are used.

Thanks to two different production mechanisms, Pyrolitic Graphite crystals can be obtained with different mosaicity and divided into two main families: Higly Annealed Pyrolitic Graphite crystals (HAPG) and Highly Oriented Pyrolitic Graphite (HOPG), showing lower and higher mosaicities, with consequently higher and lower resolutions, respectively. Both of these crystal types can be realised with different ad hoc geometries, making them suitable to be used in the Von Hamos configuration [17], combining the dispersion of a flat crystal with the focusing properties of cilindrically bent crystals.

Von Hamos spectrometers have been extensively used in the past providing very promising results in terms of spectral resolution [4,16,18] but all of the available works in literature report measurements for effective source dimensions of some tens of microns or parallel and monochromatic beams; these configurations are achieved either with microfocused X-ray tubes, or with a set of slits and collimators placed before the target to minimize the activated area. Possible spectrometer versions based on a full-cylinder geometry, to be used with a pixelated area detector instead of a monodimensional strip detector, are also possible; this may lead to an increase in both spectral resolution, thanks to a better integration of the bent lines produced by the crystal, and overall efficiency, thanks to a higher solid angle acceptance, but the effect of a hundred of microns source size are still to be tested [19].

In this work, we compare the response of three different thickness 206.7 mm curvature radius HAPG crystals and of two different mosaicity (HAPG and HOPG) 100 μm thickness 103.4 mm curvature radius crystals to $Cu(K_{\alpha 1,2})$ and $Fe(K_{\alpha 1,2})$ lines, respectively. The resulting peak resolutions and reflection efficiencies will be discussed.

2. Spectrometer Setup and Geometry

2.1. Von Hamos Geometry

The spectrometer configuration used in the measurements presented in this work is the "semi" Von Hamos one, in which the X-ray source and the position detector are placed on the axis of a cylindrically bent crystal (see Figure 1); this geometrical scheme allows an improvement in the reflection efficiency due to the vertical focusing. As a consequence, for each X-ray energy, the source-crystal (L_1) and the source-detector (L_2) distances are determined by the Bragg angle (θ_B) and the curvature radius of the crystal (ρ_c):

$$L_1 = \frac{\rho_c}{sin\theta_B},$$

$$(1)$$

$$L_2 = L_1 sin\phi.$$

$$(2)$$

In Figure 1, a schematic of the dispersive plane is shown where the X-ray source is sketched in orange, the HAPG/HOPG crystal in red and the position detector in blue and green for the standard and the "semi" Von Hamos configuration, respectively; this latter configuration is particularly suitable when used with strip detectors. On one hand, it allows for having a wider dynamic range, for a fixed spectrometer geometry, with respect to the standard one (for more details, see [20,21]). On the other hand, in the standard configuration, photons may impinge on the pixels almost parallel to their surface; the resulting short path through the silicon bulk of the pixel (450 μm in the Mythen2-1D case) may result in non detected photons and hence to a lower efficiency. In the figure, ρ_c is the crystal curvature radius, θ_B is the Bragg angle, $\phi = \pi - \theta_B$, L_1 is the source-crystal distance, L_2 is half of the resulting source-detector distance and θ_M is the position detector rotation angle with respect to the standard Von Hamos configuration.

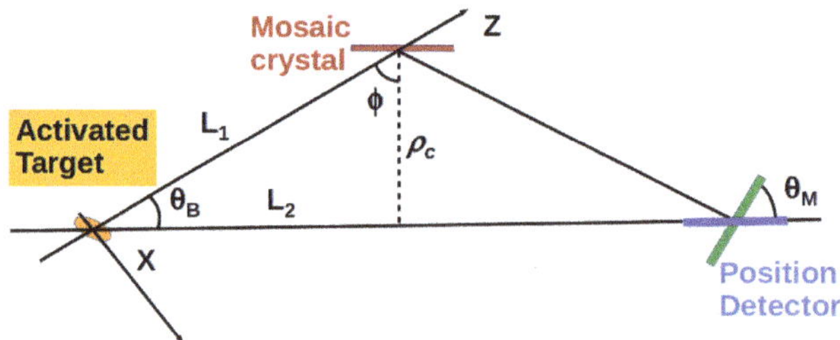

Figure 1. Von Hamos schematic geometry schematic of the dispersive plane (not in scale): the X-ray source is pictured in orange, the HAPG/HOPG crystal in red and the position detector in blue and green for the standard and the "semi" Von Hamos configuration, respectively (for more details, see [20,21] and text). On the other hand, in the standard configuration, photons may impinge on the pixels almost parallel to their surface; the resulting short path through the silicon bulk of the pixel (450 μm in the Mythen2-1D case) may result in non detected photons and hence to a lower efficiency. ρ_c is the crystal curvature radius, θ_B is the Bragg angle, $\phi = \pi - \theta_B$, θ_M is the position detector rotation angle with respect to standard VH configuration, L_1 is the source-crystal distance and L_2 is half the resulting source-detector distance.

Spectrometer Components

A thin foil (target) is placed inside an aluminum box and activated by a XTF-5011 Tungsten anode X-ray tube, produced by Oxford Instruments (systems for research, Abingdon-on-Thames, UK), placed on top of the box; the center of the foil, placed on a 45° rotated support prism, represents the origin of the reference frame in which Z is the direction of the characteristic photons emitted by the target and forming, with the *x*-axis, the Bragg reflection plane, while *y* is the vertical direction, along which primary photons generated by the tube are shot. Two adjustable motorized slits (Standa 10AOS10-1) with 1 μm resolution are placed after the 5.9 mm diameter circular exit window of the aluminum box in order to shape the outcoming X-ray beam. The various used HAPG/HOPG crystals, produced by the Optigraph company in Berlin, Germany [10], are deposited on different curvature radii Thorlabs N-BK7 30×32 mm^2 uncoated Plano-Concave Cylindrical lenses, held by a motorized mirror mount (STANDA 8MUP21-2) with a double axis $\ll 1$ *arcsec* resolution and coupled to a 0.01 μm motorized vertical translation stage (STANDA 8MVT40-13), a 4.5 *arcsec* resolution rotation stage (STANDA 8MR191-28) and two motorized 0.156 μm linear translation stages (STANDA 8MT167-25LS). The position detector is a commercial MYTHEN2-1D 640 channels strip detector produced by DECTRIS (Zurich, Switzerland), having an active area of 32×8 mm^2 and whose strip width and thickness are, respectively, 50 μm and 420 μm; the MYTHEN2-1D detector is also coupled to a positioning motorized system identical to the one for the HAPG/HOPG holder. Finally, a standar Peltierd Cell is kept on top of the strip detector in order to stabilize its temperature in the working range of 18°–28°. The resulting 10-axis motorized positioning system is mounted on a set of Drylin rails and carriers to ensure better stability and alignment and, in addition, to easily adjust source-crystal-detector positions for each energy to be measured.

2.2. Source Size in Dispersion Plane

As pointed out in Section 1, the actual limitations on the possible usage of crystal spectrometers for extended targets are represented by the requirement of a point-like source; however, using a pair of slits as the one described in the previous section, it is possible to shape the beam of X-rays emitted by an extended and diffused target in such a way as to simulate a virtual point-like source.

Referring to Figure 2, this configuration is obtained setting the position (z_1 and z_2) and the aperture (S_1 and S_2) of each slit in order to create a virtual source between the two slits (z_f, green solid lines on Figure 2), an angular acceptance $\Delta\theta'$, and an effective source S_0' (green) which can be, in principle, as wide as necessary. The $\Delta\theta'$ angular acceptance could also be set to any value, provided it is large enough to ensure that all the θ_B corresponding to the lines to be measured are included.

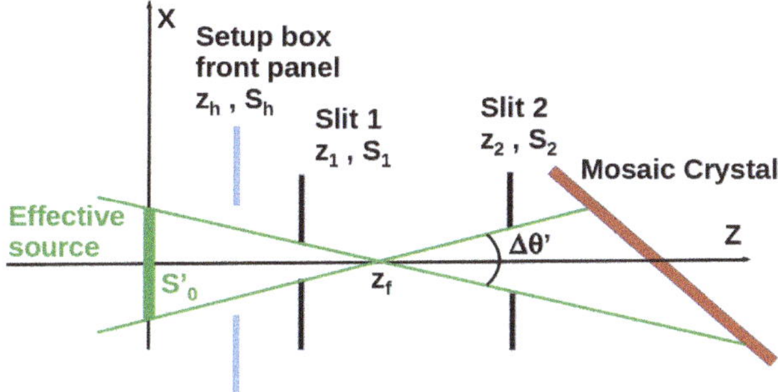

Figure 2. Beam geometry on the dispersive plane (not in scale): the position (z_1 and z_2) and the aperture (S_1 and S_2) of two slits are used to create a virtual source between the two slits (z_f, green solid lines), and effective source S_0' (green) and an angular acceptance $\Delta\theta'$; the HAPG/HOPG crystal is pictured in red, the two slits are shown in black while z_h and S_h are the position and the diameter of the circular exit window of the aluminum box front panel (light blue), respectively.

For the following discussion, we call the photons leaving the target from the central position and meeting the Bragg rule "nominal"; since the photons are isotropically emitted from the whole target foil, some of them may have the correct energy and angle to be reflected but originate from a point of the target near the nominal one. As far as this mislocation is below the limit given by the mosaic spread of the crystal, such photons are also reflected under the signal peak worsening the spectral resolution; on the contrary, when this mislocation exceeds this limit, these photons are reflected outside the signal peak. In the same way, photons not emitted in parallel to the nominal ones may still impinge on the HAPG/HOPG crystals with an angle below its mosaic spread and be then reflected under the signal peak also affecting the spectral resolution. On the contrary, if the impinging angle is out of this limit, those photons are not reflected on the position detector. As a consequence, for each energy, there is the possibility to find the right slits configuration leading at the maximum source size keeping the resolution below the desired limit.

For each chosen $\Delta\theta'$, S_0' pair, the corresponding values of the slits' apertures S_1 and S_2 can be found; first, we define the position of the intersection point z_f:

$$z_f = \frac{S_0'}{2} ctg \frac{\Delta\theta'}{2}.$$ (3)

Then, the two slits' apertures and the vertical illuminated region of the HAPG/HOPG are defined by:

$$S_1 = \frac{z_f - z_1}{z_f} S_0',$$

$$(4)$$

$$S_2 = \frac{z_2 - z_f}{z_f} S_0'.$$

$$(5)$$

2.3. Possible Effects of the Crystal Thickness

As introduced in Section 1, for a given crystal mosaicity, the graphite thickness has a double effect; on one side, increasing the thickness of the crystal leads to a higher reflectivity because of a higher probability to find a properly oriented crystallite; on the other side, photons reflected at different depths result slightly shifted on the position detector. In fact, two photons with exactly the same angle and energy would be both reflected from the same microcrystallite (if coming from the same point, so following the exact same path), but two photons with a slightly different energy (or angle) may penetrate more in the crystal before being reflected by a properly oriented microcrystallite. This introduces an additional and bigger misplacement on the position detector with respect to the one only due to the mosaicity. The situation is shown in the schematic of Figure 3: photons emitted from different part of the source (orange spot) with the same λ_2, θ_2 can be reflected at different depths in the crystal (red line) by two distinct crystallites (blue rectangulars), resulting in different positions x_2 (orange dotted line) and x_2' (black solid line) with respect to the nominal one with λ_1, θ_1 (x_1, solid orange line).

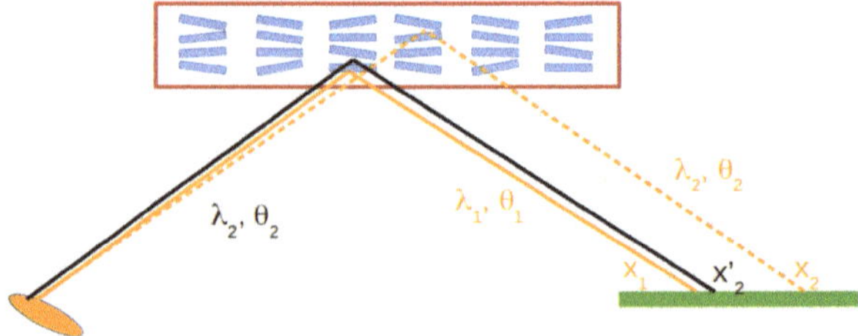

Figure 3. Depth reflection schematic: photons emitted from different part of the source (orange spot) with the same λ_2, θ_2 can be reflected at different depths in the crystal (red line) by two distinct crystallites (blue rectangulars), resulting in different positions x_2 (orange dotted line) and x_2' (black solid line) with respect to the nominal one with λ_1, θ_1 (x_1, solid orange line).

The above mentioned aberration effect is well known and studied (see for example [12]); however, it is interesting to check whether the worsening of the resolution due to the crystal thickness is still a dominant effect also when the effective source size is about 1 mm or if instead the broadening induced by the source size becomes more important.

3. Results

In this section, we present the spectra obtained for the Cu($K_{\alpha1,2}$) and Fe($K_{\alpha1,2}$) lines (see Table 1); for each measurement, the corresponding geometrical parameters are listed in Table 2, where θ_B^{set} is the central Bragg angle value used for the calculation. The crystal curvature radius is chosen in order to make a compromise between the energy resolution and the signal rate, since higher ρ_c leads to longer paths meaning better resolution but higher X-ray absorption from the air. Slits' positions z_1 and z_2 are chosen such as to have a vertical dispersion at the HAPG/HOPG position smaller than the crystal size (30 mm). The typical fitting function for peaks obtained with micrometric sources is a Lorentzian, which can be used to take into account the natural lineshape of the atomic transitions and, in case, the mosaic spread of a crystal; however, in our analysis, we carefully checked our spectra and fitted them either with a Lorentzian, a Gaussian, and a convolution of the 2 (Voigt function), and we concluded that, in our case, in which the dimension of the source is not a few microns anymore, the Gaussian shape better reproduces the broadening induced by the bigger source size. In the following subsections, the presented spectra are then fitted with Gaussian functions; for each acquired spectra, since the angular separation between Cu and Fe $K_{\alpha1,2}$ lines is very small ($\Delta\theta_{1,2}(Cu) = 0.034°$, $\Delta\theta_{1,2}(Fe) = 0.035°$), a linearity regime is assumed between the pixel position on the strip detector and the X-ray energy. The calibration is then obtained by a linear interpolation of the peak mean values obtained by the fits of the position spectra; this linear function is then used to convert position spectra into energy spectra. For each plot, the rate reported in the yellow box is calculated as the sum of the integrals of the $K_{\alpha1,2}$ peaks divided by the data acquisition (DAQ) time (1 h DAQ for all the reported measurement).

Table 1. List of the X-ray lines used in this work and the corresponding Bragg angles θ_B.

Line	E (eV)	θ_B (°)
Fe($K_{\alpha1}$)	6403.84	16.774
Fe($K_{\alpha2}$)	6390.84	16.809
Cu($K_{\alpha1}$)	8047.78	13.276
Cu($K_{\alpha2}$)	8027.83	13.310

Table 2. List of the measurements presented in this work and their main beam parameters.

Line	θ_B^{set} (°)	ρ_c (mm)	L_1 (mm)	L_2 (mm)	z_1 (mm)	z_2 (mm)
Fe($K_{\alpha1,2}$)	16.792	103.4	358.46	343.15	76	257
Cu($K_{\alpha1,2}$)	13.293	206.7	900.54	876.33	60	820

3.1. Reflection Efficiency and Resolution for Different HAPG Thickness

We present the results obtained from several measurements of the Cu($K_{\alpha1,2}$) lines, using three different 206.7 mm curvature radius HAPG crystal of 20 μm, 40 μm, and 100 μm thickness, all having a mosaicity, measured and declared by the producer, of $\omega_{FWHM} = 0.09° \pm 0.015°$. As an example, we show the comparison spectra obtained for each crystal for $\Delta\theta' = 0.5°$ and with effective source sizes of 1000 μm in Figure 4, while, for the other source sizes, the results are summarized in Table 3. In the figure, the three spectra for the 20 μm, 40 μm and 100 μm cases are shown in the upper, middle and lower panels, respectively. For each measurement, the data taking time is one hour and the fitting function is a double Gaussian with common σ for the Cu lines and a polynomial for the background.

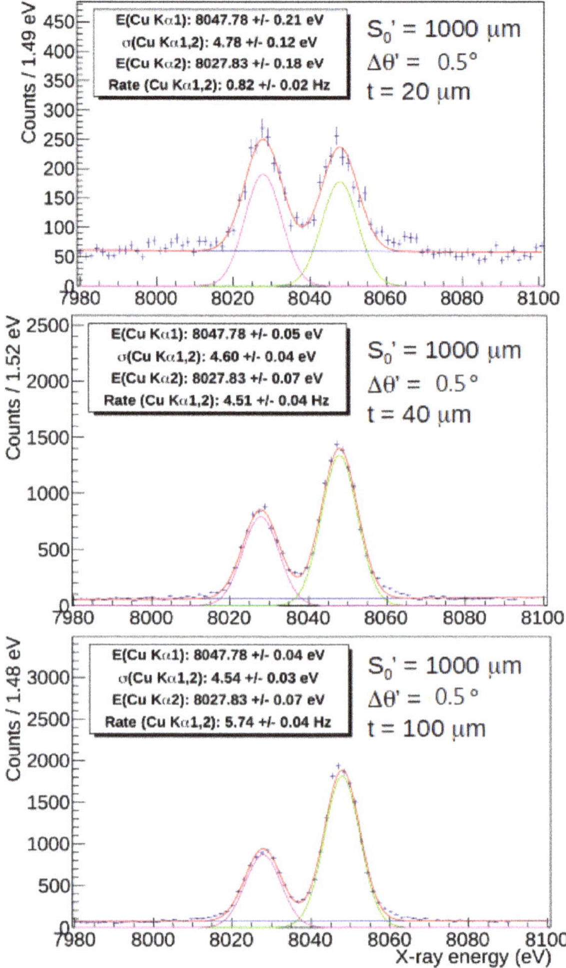

Figure 4. Fitted spectra of $Cu(K_{\alpha1,2})$ lines for $S'_0 = 1000$ μm and $\Delta\theta' = 0.5°$: $K_{\alpha1}$, $K_{\alpha2}$, polynomial background and total fitting functions correspond to the green, violet, blue and red curves, respectively, while t in the text box refers to the thickness of the HAPG crystal.

3.2. Reflection Efficiency and Resolution for Different Crystal Mosaicities

An effect similar to the one due to the crystal thickness is played by the mosaicity: for a given thickness, an increase in the mosaicity leads to a higher reflectivity because of a higher probability to find a properly oriented crystallite but, as a drawback, a defocusing effect in the reflection plane is induced [4]. We present the results obtained from several measurements of the $Fe(K_{\alpha1,2})$ lines, using two different 103.4 mm curvature radius crystal of mosaicity $\omega_{FWHM} = 0.11° \pm 0.01°$ (HAPG) and $\omega_{FWHM} = 0.45° \pm 0.03°$ (HOPG); for each one, the data taking time is one hour and the fitting function is a double Gaussian with common σ for the Cu lines and a polynomial for the background. Unlike the measurements in Section 3.1, in this case, it was not possible to obtain spectra for the same $\Delta\theta'$ for each S_0' value because of the geometrical limitations imposed by the Von Hamos configuration. For example, to have $S_0' = 400$ μm and $\Delta\theta' = 0.3°$, slit S_1 aperture should be only 2 μm, while to have $S_0' = 800$ μm and $\Delta\theta' = 0.18°$, slit S_2 aperture should be only 7 μm; for both values, not only is the rate not sufficient to acquire meaningful 1 h data taking spectra, but they are also below the error on the $\Delta\theta'$ induced by the x and z position of the slits.

As an example, the comparison spectra obtained for the two type of crystal for $S_0' = 500\,\mu$ and $\Delta\theta' = 0.18°$ combination are presented in Figure 5, where the HAPG and HOPG spectra are shown in the upper and lower panels, respectively. The results obtained for difference $S_0'\,\Delta\theta'$ combinations are reported in Table 3.

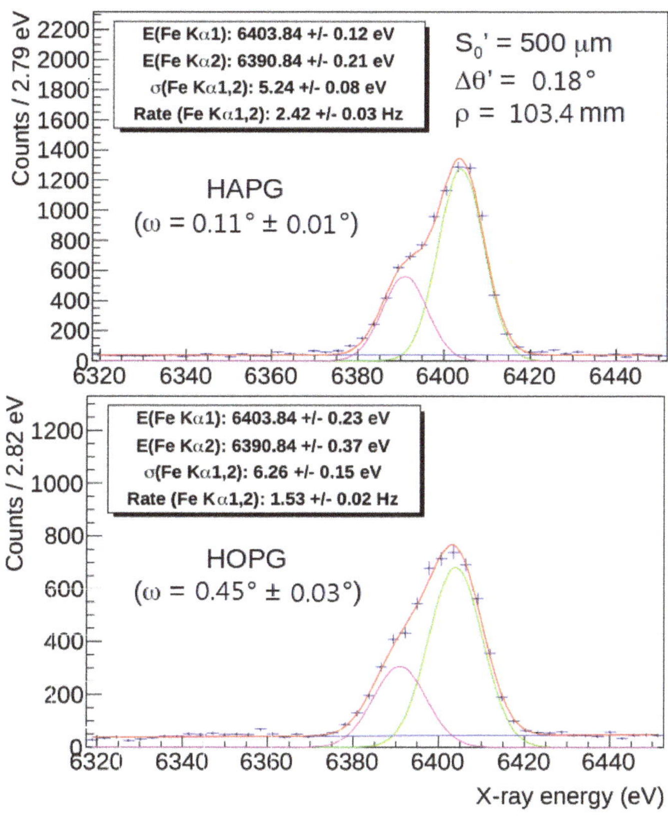

Figure 5. Fitted spectra of Fe($K_{\alpha 1,2}$) lines for $S_0' = 500\,\mu$m and $\Delta\theta' = 0.18°$: $K_{\alpha 1}$, $K_{\alpha 2}$, polynomial background and total fitting functions correspond to the green, violet, blue and red curves, respectively.

The obtained peak precisions, resolutions and total rates obtained in the different measurements are summarized in Table 3.

Table 3. Summary of the peak precisions, resolutions and total rates obtained in the different measurements.

Line	ω_{FWHM} (°)	S'_0 (µm)	$\Delta\theta'$ (°)	Thick (µm)	$\delta E_{\alpha 1}$ (eV)	$\delta E_{\alpha 2}$ (eV)	$\sigma(K_{\alpha 1,2})$ (eV)	$R(K_{\alpha 1,2})$ (Hz)
Cu($K_{\alpha 1,2}$)	0.09 ± 0.015	800	0.5	20	0.24	0.24	3.64 ± 0.24	0.31 ± 0.01
Cu($K_{\alpha 1,2}$)	0.09 ± 0.015	800	0.5	40	0.06	0.08	3.79 ± 0.05	2.42 ± 0.04
Cu($K_{\alpha 1,2}$)	0.09 ± 0.015	800	0.5	100	0.05	0.08	3.79 ± 0.04	3.05 ± 0.03
Cu($K_{\alpha 1,2}$)	0.09 ± 0.015	900	0.5	20	0.22	0.18	4.39 ± 0.13	0.66 ± 0.01
Cu($K_{\alpha 1,2}$)	0.09 ± 0.015	900	0.5	40	0.05	0.08	4.18 ± 0.04	3.49 ± 0.03
Cu($K_{\alpha 1,2}$)	0.09 ± 0.015	900	0.5	100	0.04	0.07	4.19 ± 0.03	4.58 ± 0.04
Cu($K_{\alpha 1,2}$)	0.09 ± 0.015	1000	0.5	20	0.21	0.18	4.78 ± 0.12	0.82 ± 0.02
Cu($K_{\alpha 1,2}$)	0.09 ± 0.015	1000	0.5	40	0.05	0.07	4.60 ± 0.04	4.51 ± 0.04
Cu($K_{\alpha 1,2}$)	0.09 ± 0.015	1000	0.5	100	0.04	0.07	4.54 ± 0.03	5.74 ± 0.04
Cu($K_{\alpha 1,2}$)	0.09 ± 0.015	1100	0.5	20	0.21	0.17	5.29 ± 0.11	1.07 ± 0.02
Cu($K_{\alpha 1,2}$)	0.09 ± 0.015	1100	0.5	40	0.05	0.08	4.93 ± 0.04	5.41 ± 0.04
Cu($K_{\alpha 1,2}$)	0.09 ± 0.015	1100	0.5	100	0.04	0.07	5.03 ± 0.03	7.18 ± 0.04
Cu($K_{\alpha 1,2}$)	0.09 ± 0.015	1200	0.5	20	0.23	0.20	6.01 ± 0.14	1.35 ± 0.02
Cu($K_{\alpha 1,2}$)	0.09 ± 0.015	1200	0.5	40	0.06	0.08	5.51 ± 0.04	6.35 ± 0.04
Cu($K_{\alpha 1,2}$)	0.09 ± 0.015	1200	0.5	100	0.05	0.08	5.55 ± 0.03	8.46 ± 0.05
Fe($K_{\alpha 1,2}$)	0.11 ± 0.01	400	0.16	100	0.13	0.22	4.82 ± 0.1	1.36 ± 0.02
Fe($K_{\alpha 1,2}$)	0.45 ± 0.03	400	0.16	100	0.26	0.41	5.76 ± 0.22	0.84 ± 0.02
Fe($K_{\alpha 1,2}$)	0.11 ± 0.01	500	0.18	100	0.13	0.22	5.24 ± 0.10	1.98 ± 0.02
Fe($K_{\alpha 1,2}$)	0.45 ± 0.03	500	0.18	100	0.23	0.36	6.06 ± 0.16	1.29 ± 0.02
Fe($K_{\alpha 1,2}$)	0.11 ± 0.01	600	0.22	100	0.13	0.23	5.84 ± 0.09	3.11 ± 0.03
Fe($K_{\alpha 1,2}$)	0.45 ± 0.03	600	0.22	100	0.31	0.37	7.30 ± 0.20	2.17 ± 0.02
Fe($K_{\alpha 1,2}$)	0.11 ± 0.01	700	0.28	100	0.10	0.16	5.79 ± 0.07	5.01 ± 0.04
Fe($K_{\alpha 1,2}$)	0.45 ± 0.03	700	0.28	100	0.25	0.35	7.36 ± 0.16	3.57 ± 0.02
Fe($K_{\alpha 1,2}$)	0.11 ± 0.01	800	0.36	100	0.11	0.16	6.29 ± 0.07	7.04 ± 0.04
Fe($K_{\alpha 1,2}$)	0.45 ± 0.03	800	0.36	100	0.21	0.29	7.43 ± 0.12	5.49 ± 0.04

4. Discussion

The different thickness measurements show that, for a given S'_0 effective source size, the resolution worsening effect induced by the crystal thickness is not predominant anymore; on the contrary, the resolution broadening induced by the source size is more important. This is a very important result to be taken into account when X-rays emitted from extended isotropic sources have to be measured. It has to be noticed that the correct ratio between $K_{\alpha 1}$ and $K_{\alpha 2}$, both in Copper and Iron, is 100:51; this ratio is somehow reversed in the spectra obtained with thin crystals (see Figure 4, top pad). This effect may be better explained using Figures 1 and 2. The setup is prepared in order to have a nominal Bragg angle (θ_B in Figure 1) tuned between the $K\alpha_1$ and $K\alpha_2$ peaks, around which the angular acceptance $\Delta\theta'$ of Figure 2 is defined. If the crystal is a bit mispositioned (not centered) with respect to this nominal direction, it could result in a setup more aligned around one of the two peaks, in this case around $K\alpha_2$. Consequently, the probability to find properly oriented microcrystallites for $K\alpha_1$ is lowered.

Concerning the mosaicity influence on the reflectivity and the resolution, a more detailed discussion can be carried out starting from the measurements. From the spectra, it is evident how, with the higher mosaicity of the HOPG crystal with respect to the HAPG, one causes a peak broadening, leading, in some cases, even to a non-separation of the two peaks, consistent with the expectation; on the contrary, the reason why there is not a corresponding increase in the measured rate is not intuitive and needs to be more carefully motivated.

We can start from the assumption that, for the same crystal thickness and dimensions, the amount of inner crystallites is always of the same order of magnitude regardless of the mosaicity value; then, according to what is available in literature [12–16] and what is declared by the crystal producers [10], the orientation distribution can be described with a Lorentzian, normalized to unity, of the form:

$$L(\theta) = \frac{1}{\pi} \frac{\Gamma/2}{(\theta^2 + \frac{\Gamma^2}{4})},$$

where Γ is the crystal mosaicity ω_{FWHM}. For each $\Delta\theta'$ value (see Figure 2), for a single photon energy, the number of the properly oriented crystallites is given by:

$$N_0 = \int_{0.5\Delta\theta'}^{-0.5\Delta\theta'} L(\theta)d\theta.$$

For each measured spectrum, almost the 100% of the $Fe(K_{\alpha1,2})$ photons will be in the interval $E_1 \leq E \leq E_2$ being $E_1 = Fe(K_{\alpha2}) - 5\sigma$ and $E_2 = Fe(K_{\alpha1}) + 5\sigma$, where now σ is the Gaussian peak resolution obtained from the fits; this will correspond to a θ interval of $\alpha = \theta_2 - \theta_1$ being $\theta_2 = \sin^{-1}\left(\frac{C_{\lambda eV}}{E_2 2d}\right)$ and $\theta_1 = \sin^{-1}\left(\frac{C_{\lambda eV}}{E_1 2d}\right)$, where d is the graphite lattice parameter and $C_{\lambda eV}$ is the $\text{Å} \rightarrow eV$ conversion factor. The final number of properly oriented crystallites will then be:

$$N = \int_{0.5\Delta\theta'+\alpha}^{-0.5\Delta\theta'-\alpha} L(\theta)d\theta.$$

The situation is shown in Figure 6, where the orientation distributions for the HAPG and HOPG are shown in each panel in red and black, respectively. The distributions are peaked around $\Delta\theta = 0°$, which represents the nominal Bragg angle; the integrals of the two Gaussians are then reported for various intervals and show how the ratio between the HOPG and HAPG number of properly oriented crystallite is changing. As a consequence, in the case of the $Fe(K_{\alpha1,2})$ lines, which are only 13 eV or 0.035° distant, the expected ratio is actually matching the expected values.

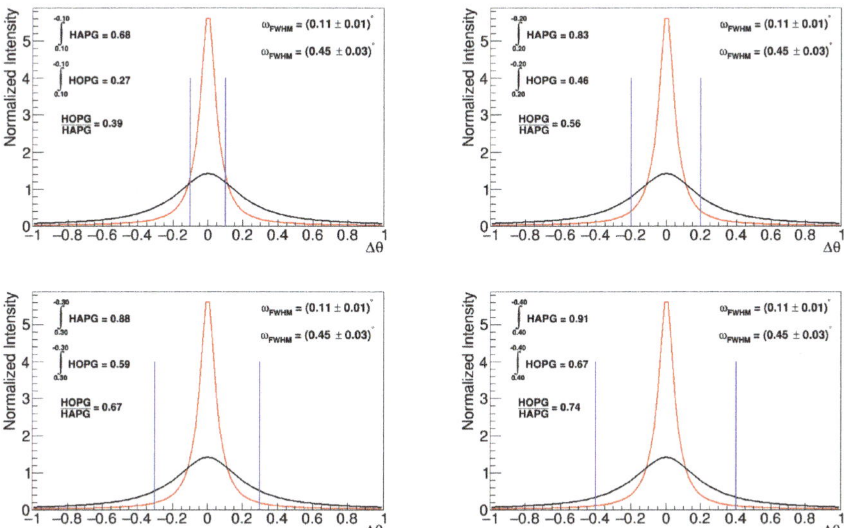

Figure 6. Orientation distributions for the HAPG (red) and HOPG (black) crystals: the distributions are peaked around $\Delta\theta = 0°$, which represents the nominal Bragg angle; the integrals of the two Gaussians are then reported for various intervals and show how the ratio between the HOPG and HAPG number of properly oriented crystallite is changing.

This description is reproducing the experimental data well, as shown in Figure 7, where, in the upper panel, the measured ratios for each $S_0' \Delta\theta'$ pairs are shown, while, in the middle panel, the corresponding calculated integral ratios are reported. In the bottom pad, the difference between the previous two is reported to show how much the calculated values differ from the measured ones. The errors on the measured points are obtained propagating the errors on the peak fitting parameters, while the errors on the calculated ones take into account the errors on the crystal mosaicities declared by the producer. The errors are then propagated in the difference plot.

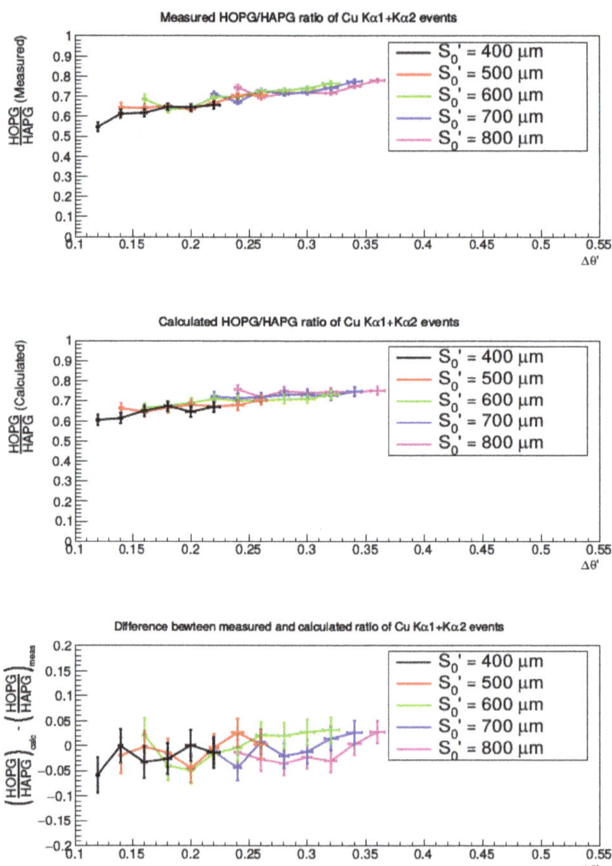

Figure 7. Comparison of the measured (**top**) and calculated (**middle**) HOPG/HAPG rate ratios; subtraction of the two graphs is shown in the bottom pad. Each point of the upper and middle plots represents the sum of both $K_{\alpha 1}$ and $K_{\alpha 2}$ number of events for the HOPG crystal divided by the same quantity for the HAPG one.

5. Conclusions

In this work, we presented the measurements of the Cu(K$_{\alpha1,2}$) and the Fe(K$_{\alpha1,2}$) lines performed with different mosaicity, thickness and curvature radius crystals in order to investigate the influence of the above mentioned parameters on the peak resolution and on the reflection efficiency. The different thickness measurements show that, for a given S_0' effective source size, the resolution worsening effect induced by the crystal thickness is not predominant anymore with respect to the effect induced by the wider source size; therefore, higher thickness values ensure higher reflectivities leading to more precise determinations of the peak positions. This is a very important result to be taken into account when X-rays emitted from extended isotropic sources have to be measured.

Concerning the mosaicity influence on the reflectivity and the resolution, a more detailed discussion was carried out starting from the measurements. On one side, we confirmed that the higher mosaicity of the HOPG crystal with respect to the HAPG one causes a peak broadening, leading in some cases even to a non-separation of the two peaks; on the other side, we measured a higher number of reflected photons using the HAPG crystal with respect to using the HOPG one. We provided a semi-quantitative explanation of this behaviour, which is in good agreement with the measured ratios.

The proposed spectrometer shows promising results for future experiments in which, like for example in the case of exotic atoms' transitions measurements, X-rays are emitted from a wide and diffused source. In these kinds of experiments, for a given spectrometer geometry, a proper choice of the crystal in terms of thickness and mosaicity becomes a crucial parameter to be set based on the main goals of the measurement and on the experimental conditions, like, for instance, the source emission rate and the solid angle acceptance. The correct combination of crystal parameters may in fact be chosen in order to obtain the desired combination of peak resolution and precision position determination. In the future, the possibility to use pixelated area detectors in a full-cylinder crystal configuration like the one mentioned in the Introduction, instead of linear strip detectors, will be also investigated; this may lead to an increase in both spectral resolution, thanks to a better integration of the bent lines produced by the crystal, and overall efficiency, thanks to a higher solid angle acceptance.

Author Contributions: Conceptualization, A.S.; methodology, A.S.; software, A.S.; formal analysis, A.S.; investigation, A.S.; data curation, A.S. and M.M.; resources, A.S., M.M. and F.S.; writing—original draft preparation, A.S.; writing—review and editing, A.S.; visualization, A.S.; supervision, A.S., C.C., and J.Z.; project administration, A.S.; funding acquisition, A.S. and J.Z.

Funding: This research was funded by the 5th National Scientific Committee of INFN in the framework of the Young Researcher Grant 2015, n. 17367/2015.

Acknowledgments: We thank the Laboratori Nazionali di Frascati INFN (LNF) and the Stefan-Meyer-Institut für subatomare Physik (SMI) staff, in particular the LNF Servizio Progettazione e Costruzione Meccanica (SPCM) service and Doris Pristauz-Telsnigg, for the support in the preparation of the setup.

Conflicts of Interest: The authors declare no conflict of interest.

References

1. Bazzi, M.; Beer, G.; Bombelli, L.; Bragadireanu, A.M.; Cargnelli, M.; Corradi, G.; Curceanu, C.; d'Uffizi, A.; Fiorini, C.; Frizzi, T.; et al. A New Measurement of Kaonic Hydrogen X-rays. *Phys. Lett.* **2011**, *B704*, 113–117. [CrossRef]
2. Gallo, A.; Alesini, D.; Biagini, M.E.; Biscari, C.; Boni, R.; Boscolo, M.; Buonomo, B.; Clozza, A.; Delle Monache G.; Di Pasquale, E.; et al. DAFNE status report. *Conf. Proc.* **2006**, *C060626*, 604–606.
3. Doriese, W.B.; Abbamonte, P.; Alpert, B.K.; Bennett, D.A.; Denison, E.V.; Fang, Y.; Fischer, D.A.; Fitzgerald, C.P.; Fowler, J.W.; Gard1, J.D.; et al. A practical superconducting-microcalorimeter X-ray spectrometer for beamline and laboratory science. *Rev. Sci. Instrum.* **2017**, *88*, 053108. [CrossRef] [PubMed]

4.   Legall, H.; Stiel, H.; Antonov, A.; Grigorieva, I.; Arkadiev, V.; Bjeoumikhov, A. A new generation of X-ray optics based on pyrolitic graphite. In Proceedings of the FEL 2006, Berlin, Germany, 27 August–1 September 2006; p. 798.

5.   Barnsley, R.; Peacock, N.; Dunn, J.; Melnick, I.; Coffey, I.; Rainnie, J.; Tarbutt, M.; Nelms, N. Versatile high resolution crystal spectrometer with X-ray charge coupled device detector. *Rev. Sci. Instrum.* **2003**, *74*, 2388. [CrossRef]

6.   Anagnostopoulos, D.; Biri, S.; Borchert, G.; Breunlich, W.; Cargnelli, M.; Egger, J.-P.; Fuhrmann, H.; Gotta, D.; Giersch, M.; Gruber, A.; et al. The Pionic Hydrogen Experiment at PSI. *Hyperfine Interact.* **2001**, *138*, 131–140. [CrossRef]

7.   Trassinelli, M.; Anagnostopoulos, D.; Borchert, G.; Dax, A.; Egger, J.-P.; Gotta, D.; Hennebach, M.; Indelicato, P.; Liu, Y.-W.; Manil, B.; et al. Measurement of the charged pion mass using X-ray spectroscopy of exotic atoms. *Phys. Lett.* **2016**, *B759*, 583–588. [CrossRef]

8.   Beer, G.; Bragadireanu, A.; Breunlich, W.; Cargnelli, M.; Curceanu, C.; Egger, J.-P.; Fuhrmann, H.; Guaraldo, C.; Giersch, M.; Iliescu, M.; et al. A new method to obtain a precise value of the mass of the charged kaon. *Phys. Lett.* **2002**, *B535*, 52–58. [CrossRef]

9.   Antonov, A.A.; Baryshev, V.; Grigoryeva, I.; Kulipanov, G.; Terekhov, Y.V.; Shipkov, N. First results on application of short-focus monochromators from formed pyrolytic graphite for X-ray fluorescent analysis using synchrotron radiation. *Rev. Sci. Instrum.* **1989**, *60*, 2462–2463. [CrossRef]

10.   Optigraph References. Available online: http://www.optigraph.eu/references.html#ref_07 (accessed on 3 April 2019).

11.   Del Rio, M.S.; Gambaccini, M.; Pareschi, G.; Taibi, A.; Tuffanelli, A.; Freund, A.K. Focusing properties of mosaic crystals. *Proc. SPIE* **1998**, *3448*, 246–255.

12.   Gerlach, M.; Anklamm, L.; Antonov, A.; Grigorieva, I.; Holfelder, I.; Kanngießer, B.; Legall, H.; Malzer, W.; Schlesiger, C.; Beckhoff, B. Characterization of HAPG mosaic crystals using synchrotron radiation. *J. Appl. Cryst.* **2015**, *48*, 1381–1390. [CrossRef]

13.   Ice, G.; Sparks, C. Mosaic crystal X-ray spectrometer to resolve inelastic background from anomalous scattering experiments. *Nucl. Instrum. Methods Phys. Res. A* **1990**, *291*, 110–116. [CrossRef]

14.   Legall, H.; Stiel, H.; Arkadiev, V.; Bjeoumikhov, A. High spectral resolution X-ray optics with highly oriented pyrolytic graphite. *Opt. Express* **2006**, *14*, 4570–4576. [CrossRef] [PubMed]

15.   Zastrau, U.; Brown, C.; Döppner, T.; Glenzer, S.; Gregori, G.; Lee, H.; Marschner, H.; Toleikis, S.; Wehrhan, O.; Förster, E. Focal aberrations of large-aperture HOPG von-Hamos X-ray spectrometers. *JINST* **2012**, *7*, P09015. [CrossRef]

16.   Zastrau, U.; Woldegeorgis, A.; Förster, E.; Loetzsch, R.; Marschner, H.; Uschmann, I. Characterization of strongly-bent HAPG crystals for von-Hámos X-ray spectrographs. *JINST* **2013**, *8*, P10006. [CrossRef]

17.   Von Hamos, L.V. Röentgenspektroskopie und Abbildung mittels gekrümmter Kristallreflektoren. I. Geometrisch-optische Betrachtungen. *Ann. Physik* **1933**, *409*, 716. [CrossRef]

18.   Shevelko, A.P.; Kasyanov, Y.S.; Yakushev, O.F.; Knight, L.V. Compact focusing von Hamos spectrometer for quantitative X-ray spectroscopy. *Rev. Sci. Instrum.* **2002**, *73*, 3458. [CrossRef]

19.   Anklamm, L.; Anklamm, L.; Schlesiger, C.; Malzer, W.; Grötzsch, D.; Neitzel, M.; Kanngießer, B. A novel von Hamos spectrometer for efficient X-ray emission spectroscopy in the laboratory. *Rev. Sci. Instrum.* **2014**, *85*, 053110. [CrossRef] [PubMed]

20.   Scordo, A.; Shi, H.; Curceanu, C.; Miliucci, M.; Sirghi, F.; Zmeskal, J. VOXES, a new high resolution X-ray spectrometer for low yield measurements with diffused sources. *Acta Phys. Polon.* **2017**, *B48*, 1715. [CrossRef]

21.   Scordo, A.; Curceanu, C.; Miliucci, M.; Sirghi, F.; Zmeskal, J. Development of a compact HAPG crystal Von Hamos X-ray spectrometer for extended and diffused sources. *arXiv* **2019**, arXiv:1903.02826.

Article

Graphite Optics—Current Opportunities, Properties and Limits

Inna Grigorieva *, Alexander Antonov and Gennadi Gudi

Optigraph GmbH, Rudower Chaussee 29, 12489 Berlin, Germany; a.antonov@optigraph.eu (A.A.); gudi@optigraph.eu (G.G.)
* Correspondence: inna.grigorieva@optigraph.eu; Tel.: +49-(0)30-6392-6570

Received: 21 November 2018; Accepted: 21 January 2019; Published: 24 January 2019

Abstract: X-ray graphite optics consists of thin layers of Pyrolytic Graphite (PG) attached to a substrate of focusing shape. Pyrolytic Graphite is a perfect artificial graphite obtained by annealing of carbon deposit at temperatures about 3000 °C under deformation. By varying the annealing conditions, one could get PG of different mosaic structure and mechanical properties. A wide variability of the reflecting layer characteristics and optics shape makes the graphite optics useful in an extended range of applications. The optics could be adjusted to applications that require moderate resolution as EDXRF (energy dispersive X-Ray fluorescence) and as well as for high-resolution applications as EXAFS (extended X-ray absorption fine structure), XANES (X-ray absorption near-edge structure) and XES (X-ray emission spectroscopy). To realize the optics with theoretically optimized parameters the relationship between the production procedure and the mosaicity and reflectivity of the optics was experimentally studied. The influence of thickness, the type of PG (Highly Oriented PG (HOPG) or Highly Annealed PG (HAPG)) and substrate characteristics on the optics performance is presented.

Keywords: HOPG; HAPG; Pyrolytic Graphite; von Hamos; mosaic spread; mosaicity; rocking curve

1. Introduction

The recent interest in the fine structure of spectra [1], low probability events [2] and low concentrations [3] accompanied with the requirement for decreased X-ray environmental impact of the source generates the needs in efficient X-ray optics. Capillary and multilayers, ideal crystals and many other devices are in use, each has its advantages and restrictions. Graphite Optics (GrO)—one of the brightest variants of X-ray optics becomes more and more popular, especially since Highly Annealed Pyrolytic Graphite (HAPG) with mosaic spread of 0.1° was created.

Due to the wide variability of the reflecting layer and of the optics shape, the GrO could be used in enlarged number of applications. The optics is helpful for applications requiring moderate energy resolution as well as for high-resolution applications.

A typical application where GrO is in use for more than 20 years is EDXRF (energy dispersive X-Ray fluorescence). The optics serves as monochromator and focusing device for the primary beam as well as a broadband filter between sample and detector. GrO as a broadband filter is essential for the detection of trace elements in heavy matrix [4] or the element under investigation in the presence of interfering element [5]. For this application, Highly Oriented Pyrolytic Graphite (HOPG optics with moderate energy resolution and good flux is more popular; however, recently, HAPG optics with improved mosaicity and better energy resolution has also been in use. Doubly curved short focus GrO of both types is exploited in commercial set-up [6].

In high resolution applications HAPG optics is used in von Hamos geometry as an effective dispersive element for X-ray absorption and emission spectroscopy. The efficiency of spectrometers based on HAPG optics makes it possible to implement the methods usually requiring Synchrotron Radiation (SR) sources such as XES (X-ray emission spectroscopy) and XAFS (X-ray absorption fine structure) at low brilliant laboratory sources [1,7]. HAPG optics is used in the first commercial X-ray absorption spectroscopy system QuantumLeap-XAS by Sigray [8].

Plasma analysis is another application field where GrO-based spectrometers characterize the temperature, density and ionization stage of warm dense plasma [9]. High thermal and radiation stability is a key advantage of GrO for the application. The optics is able to work in high fluencies of neutrons and laser-plasma debris.

For each application, a certain form of PG could be chosen and then the characteristics of the reflecting layer could be additionally tuned.

2. Properties and Structural Peculiarities of the Reflecting Layer in Graphite Optics

Graphite Optics produced by Optigraph GmbH [10] consists of a layer of Pyrolytic Graphite (PG) deposited on a substrate of required shape. The layer could be up to a few hundred microns thick.

Pyrolytic Graphite is a perfect artificial graphite produced by annealing of Pyrolytic Carbon—unordered graphite material that is obtained by thermal cracking of carbon containing gas, mainly methane, on a hot surface (Figure 1) [11]. The material has a two-dimensional ordering of crystal cells, a clear conical structure and mosaic spread of 30°.

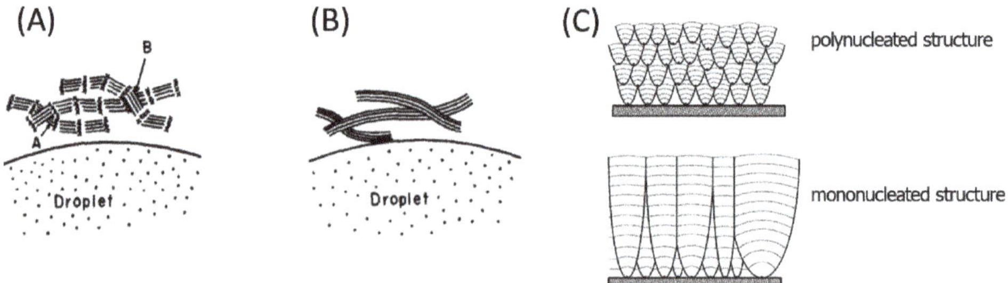

Figure 1. Production of Pyrolytic Carbon—initial material for Pyrolytic Graphite. (**A**) Before dehydrogenation; (**B**) after dehydrogenation (Reproduced with permission from [11] © Elsevier); (**C**) cone structure of pyrolytic carbon [12], modified.

Annealing at temperatures near 3000 °C orders the material structure (Figure 2). The structural changes are enhanced when annealing is accompanied by deformation. Different combinations of the annealing temperature and the deformation type lead to different structure and different forms of PG [13]. The forms with well-aligned structure are considered as graphite mosaic crystal and used as optical element for X-rays and neutrons.

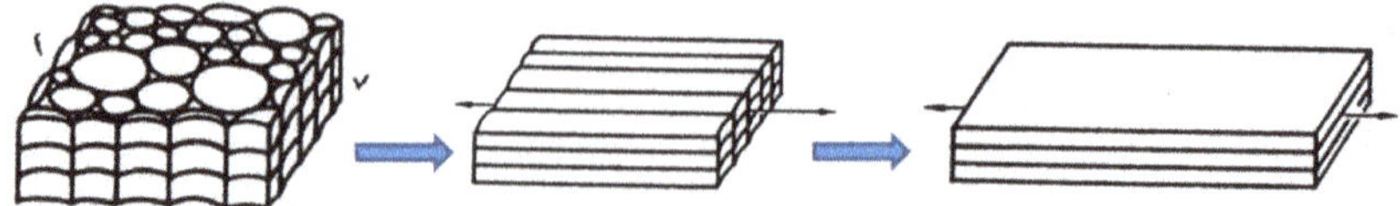

Figure 2. Schematic representation of structural changes associated with annealing under pressure (Reproduced with permission from [13] © Elsevier, modified).

Pyrolytic graphite could be considered to some approximation as a set of bulk graphene chips and has very interesting properties: it is pure carbon (99.999%) with highly anisotropic structure and properties. Its thermal conductivity in C-C plane is about 2000 W/(m·K) that is four times higher than for Cu. The material is electrically conductive along the C-C plane and is an insulator in the perpendicular direction. It can withstand high thermal and radiation loads, is chemically neutral and ecologically friendly [14].

The PG properties result in the largest integrated reflectivity of graphite among all other crystals. The dramatic increase of the integral reflectivity in comparison to ideal crystals is governed by crystal mosaicity that determines the width of the reflected energy window. Low absorption leads to the increase of the reflecting layer thickness and of the number of crystal planes involved in X-ray diffraction. As the result, the integral reflectivity of GrO is more than an order of magnitude higher than for ideal crystals [15,16]. The crystal brightness is enhanced by efficient focusing. The optics had high collection efficiency due to the possibility of being realized in any shape including full figure of revolution [5] and short focus geometries [6]. When the crystal is placed equidistant between source and detector, mosaic focusing in the detection plane provides not only additional reflectivity increase, but also high resolving power [17].

GrO works in a large energy range, including the energies higher than 10 keV [18,19]. Due to its unique brightness, the higher orders of reflection (d004, d006) could be used for widening the working energy range and increasing the crystal resolution. GrO is very robust: it endures increased temperatures, radiation and unfriendly handling.

The PG layer has a complex anisotropic structure. It contains big structural units—blocks or domains consisting of smaller crystallites [20]. The size of the blocks (domains) is a few hundreds μm along the surface and an order of magnitude less in the perpendicular direction (Figure 3) [21]. The average size of the structural units depends on the annealing procedure. The wide scatter of the results is clearly seen in review [22] where the average size of blocks and crystallites in conventional HOPG measured by different methods and different researchers were collected. We guessed that the difference in values is derived mainly from unequal annealing conditions of the samples obtained from different sources and by different procedures.

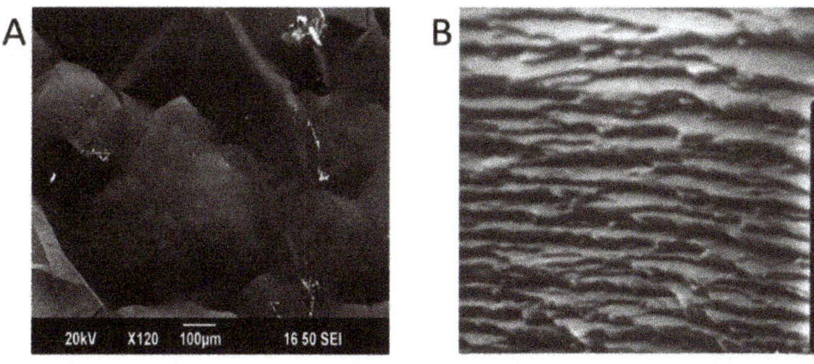

Figure 3. Structure of Pyrolytic Graphite (PG)-layers in Graphite Optics (GrO) by Scanning Electron Microscope. (**A**) Two level structure in plane: big blocks of a few hundreds μm, small crystallites of 10–30 μm; (**B**) the structure of crystal edge after laser cut: the edges of the blocks of 1–3 μm thick are clearly seen. (Reproduced with permission from [21] © Springer, modified).

Some researchers noted the existence of reflecting areas of ≤1 mm size in HAPG [23,24]. It is still not clear where these areas are blocks of increased size or there is one more structural level.

The units are well aligned along the average direction of C-C planes. The mosaicity that could be considered as a deviation of the structural unit orientation from this average plane is mainly determined by blocks; the crystallites within the block are misaligned at an angle of about 0.1°.

The variation of annealing procedure changes the material structure that in its turn determines the mosaic spread and flexibility of the material. Currently, Optigraph uses three forms of PG for optical applications, which differ by annealing history and as the result by crystal structure and characteristics they supply to the corresponding GrO.

2.1. Types of GrO

2.1.1. Optics Based on Conventional HOPG Crystal

HOPG as a monochromator and model object for fundamental research was developed in the beginning of 70 s [14]. As an optical element, HOPG was famous for its unusual brightness; however, it has moderate resolution owing to relatively big mosaic spread. Commercially available crystals have mosaic spread of 0.4° ± 0.1° or bigger. The bending of HOPG occurs at 3000 °C during annealing under pressure on concave/convex graphite pistons.

From the very beginning, there were attempts to design graphite optics on the base of flat and singly bent graphite monochromators [25]. Similarly to ideal crystals, the thin plates of HOPG were slightly bent and glued on a substrate; however the devices did not have significant efficiency in spite of the highest brightness of graphite monochromator. Until recently, this optics has some niche application for simple instruments where a moderate performance is compensated by relatively low cost.

The impossibility of the crystal bending to small radii, difficulties in production of custom and doubly bent shapes prevented appearing efficient GrO on the market.

2.1.2. HOPG Optics

The extensive research of the factors determining the PG structure and properties made at the end of last century [14,20,26,27] resulted in a novel approach to GrO that was designed on the base of thin flexible films [28].

The film flexibility is a result of a special combination of annealing and deformation that eliminate the vast defect regions along the grain boundaries existing in conventional HOPG. Figure 4 shows the grain structure of conventional and flexible HOPG of the same mosaicity by acoustic microscopy [28]. In conventional HOPG, the grain boundaries are marked by contrast zones that are not visible in the flexible form.

The film flexibility grants a chance to produce custom shaped optics at room temperature and notably decreases the production costs. Moreover, the structural changes allow attaining significantly lower crystal mosaicity than by classical technology.

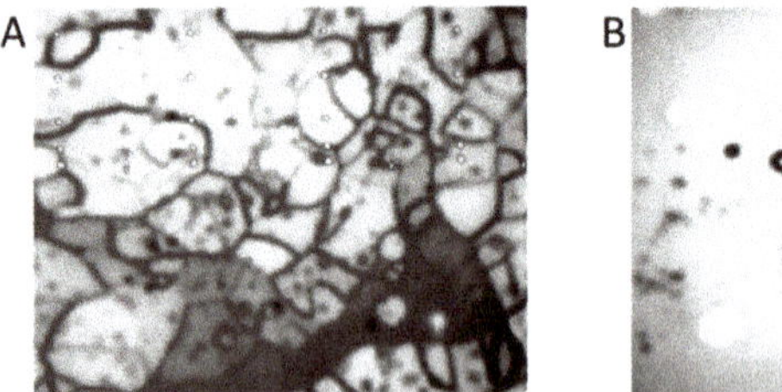
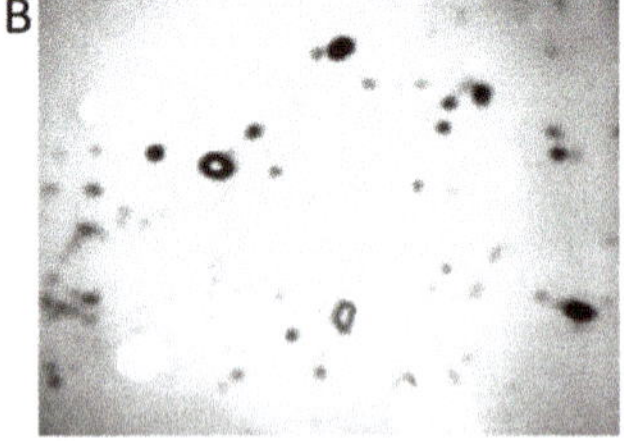

Figure 4. Structure of rigid and flexible HOPG films by acoustic microscope. (**A**) Rigid material, could be bent plastically at 3000 °C or elastically as a film of 50–200 μm; (**B**) flexible HOPG is acoustically transparent. (Reproduced with permission from [28]. © Wiley-VCH Verlag GmbH & Co. KGaA).

It is of a special importance that the resolution of bent optics does not degrade versus a flat one. The flexible films are bent plastically at room temperature, and the deformation causes neither stress, nor crucial structural changes [29].

HOPG optics is made of a set of thin films with a mosaicity of 0.1 degree. The thickness of a film could be from 5 to 20 μm. The films are split off from the material denoted as flexible HOPG. The films adhere to each other and to optically polished substrate without glue and a layer of required thickness could be deposited on a substrate of any shape at room temperature. The mosaicity of the optics increases with the number of the films involved.

HOPG optics could be produced with radii down to a few mm and as full figure of revolution. The optics offers efficient focusing with moderate resolution and relatively wide bandpass.

HOPG optics finds application mainly in the fields where the priority is the flux and not the resolution. The optics has relatively moderate requirements to the substrate and is preferable for complex short focus shapes [30,31].

Providing efficient focusing, HOPG-optics still retains moderate mosaic spread similar to the conventional HOPG. In spite of a low mosaic spread of very thin films the optics with practically useful thickness of reflecting layer does not fit the high-resolution applications.

2.1.3. HAPG Optics

The upgrade of the annealing procedure permitted to increase by an order of magnitude the thickness of flexible monofilms with mosaicity of about $0.1°$. The film adheres to optically polished substrate and has increased flexibility. As the annealing procedure significantly differs from the technology of conventional and flexible HOPG, the material got a special name HAPG.

HAPG has more perfect structure and decreased intrinsic width in comparison with both variants of HOPG. That leads to better combination of sensitivity and resolution. In von Hamos geometry the HAPG optics reaches a resolution close to ideal crystals keeping an order of magnitude higher brightness [29,32]. The direct comparison of HAPG optics at von Hamos geometry with ideal crystals such as Si [16], Ge [15] and with mosaic crystals such as LiF [16,23] demonstrates in all cases a significant integrated reflectivity gain at comparable resolving power.

Tuning the GrO to a given task one could not only choose the required form of PG layer and the optics geometry, but also vary the parameters of reflecting layer in order to achieve maximal efficiency.

The resolution and brightness of the optics both depend on the thickness and mosaicity of graphite layer; however, the trends are contrariwise. X-rays penetrate relatively deep into the graphite and reflect from the depth up to a few hundred microns depending on the energy and crystal mosaicity. Aberrations derived from the mismatch between the rays reflected from the top and the bottom of the reflecting layer limit the crystal thickness by the resolution on request. A smaller curvature radius needs a thinner crystal layer if the same resolution should be achieved.

Besides the depth broadening, there are other factors degrading the GrO energy resolution; they are thoroughly discussed in [33,34]. Increased collection solid angle does not always result in increased efficiency, because a considerable fraction of the collected photons could be lost due to aberrations; thus, the crystal with decreased mosaicity could compensate its potentially lower reflectivity by tight focusing. As a result, the optimal type of GrO, as well as the choice between ideal crystals and graphite optics, depends on the required spectral resolution and other limitations of the given application [35].

To choose the thickness of the graphite layer for the optimal ratio between resolution and brightness of the optics one needs not only theoretical estimations, but also experimental data that establish the correlation between mosaicity, reflectivity and thickness of the layer for different PG forms and deposition procedures.

The graphite film follows the substrate precisely; thus, the substrate availability determines the chance to realize optics according to the theoretical calculation. The films adhere to the optically polished substrate made from glass, quartz and some other materials. Aluminum substrates often require some gluing. However, aluminum substrates could offer more various shapes, could work under trying conditions and they are cheaper and easier to handle. Therefore, the choice of substrate material also becomes a question of compromise and experiments revealing the influence of the substrate on the optics parameters are essential.

In order to clarify these questions, we analyzed characteristics such as mosaic spread and integrated reflectivity of the graphite layers deposited on different substrates, in different ways and of different thickness.

3. Results

3.1. HAPG versus HOPG

The thickness of a single film with mosaicity of 0.1° is limited by 10–15 µm for flexible HOPG and 120 µm for HAPG. HOPG films up to 50 µm thick could still adhere to optically polished glass substrate; however, the mosaicity of such films increases up to 0.25° and the flexibility significantly decreases. Increased surface roughness of the substrate results in worse film adhesion. For the aluminum substrates used in this work, the maximal thickness of the adhering film decreased down to 30/80 µm for HOPG and HAPG films, respectively.

As films adhere not only to optically polished substrate, but also to each other, there is a possibility to increase the thickness of reflecting layer using a set of films. The approach results in a significant mosaicity increase with the number of the films involved. The required thickness could be obtained by the deposition of a few thicker films with increased mosaicity or a bigger amount of thin films with mosaicity of 0.1°. To compare both ways of deposition, the mosaicity and reflectivity of HOPG layers produced from thin films (5–10 µm each) and from films of maximal possible thickness (30–50 µm each) were measured (Figure 5).

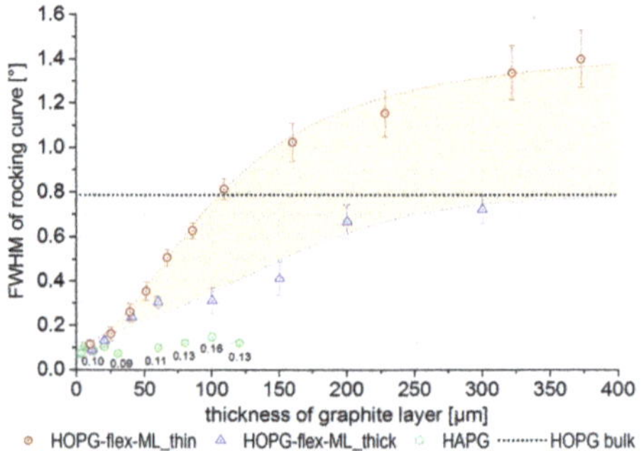

Figure 5. Mosaic spread of GrO on glass substrate depending on graphite layer thickness: red circles—HOPG layer composed of thin films (5–10 µm); blue triangles—HOPG layer composed of thick films (30–50 µm); green circles—HAPG layer applied as a single film. The error bars represent the standard deviations of every 10 measurement points.

In the "thin-film method" the film mismatch rapidly increases the mosaicity of HOPG optics up to the values higher than those of the source material—flexible HOPG (up to 1.5° in comparison to 0.8°). In the case of "thick-film method", the mosaicity increases slower and reaches the values of the source material. The HAPG layer could be applied as a single film up to 120 µm thick retaining the mosaicity of about 0.1° (Figure 6). Stacking of HAPG layers from a set of single films also leads to mosaicity similar, but slower, increase to flexible HOPG. For example, the mosaicity of 40 µm HAPG layer made of 10 films exceeded only twice the mosaicity of the monofilm variant. For convenient comparison of HOPG and HAPG the results for HAPG are also additional placed on Figures 5 and 7. The reflectivity of the HOPG layer reaches a "plateau" at 200–300 µm (Figure 7) and this thickness of the reflecting layer is typically used for HOPG optics working below 10 keV. The mosaicity of such optics lies in the range of 0.8°–1.5° (Figure 5). Due to lower mosaicity, HAPG optics has a smaller integral reflectivity versus HOPG optics, while the peak reflectivity is the same.

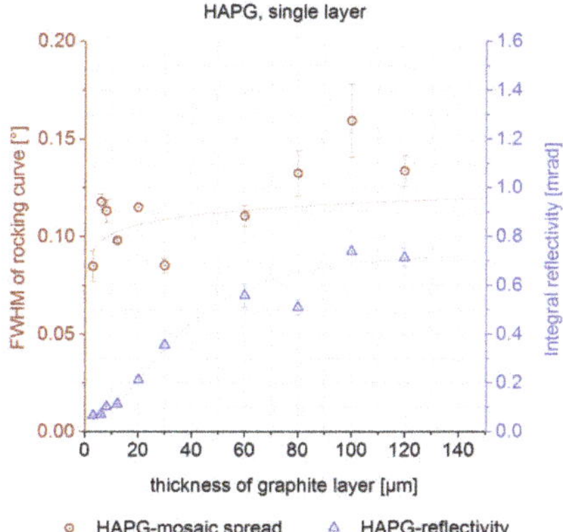

Figure 6. Mosaicity and reflectivity of HAPG deposited as a single film on glass substrate depending on the graphite layer thickness: red circles—mosaicity; blue triangles—reflectivity. The error bars represent the standard deviations of every 10 measurement points.

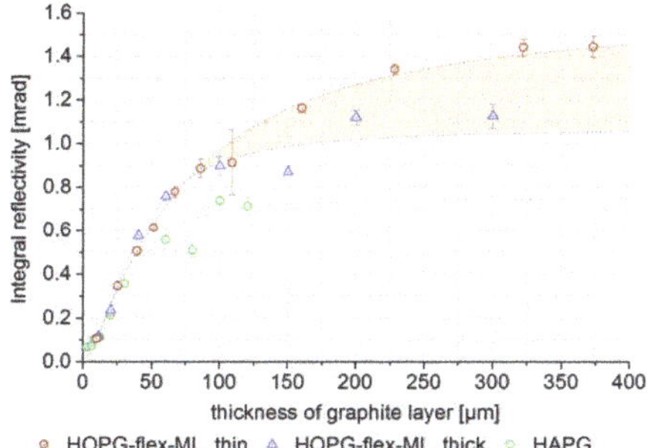

Figure 7. Reflectivity of GrO on glass substrate depending on graphite layer thickness: red circle—HOPG layer composed of thin films (5–10 µm); blue triangles—HOPG layer composed of thick films (30–50 µm); green circles—HAPG layer applied as a single film. The error bars represent the standard deviations of every 10 measurement points.

3.2. Glass Substrates versus Aluminum Substrates

In order to test the influence of the substrate materials on the characteristics of GrO, aluminum and glass were used as substrates for HAPG films of different thickness. A single film layer was deposited sequentially on two types of the substrates. Each film was deposited first on aluminum substrate of worse surface quality and afterwards it was redeposited on optically polished glass substrate. The mosaicity was measured after first and second deposition. The same HAPG film on well-polished aluminum substrate has twice higher mosaicity than on optically polished glass (Figure 8). We would like to emphasize that the film improved its mosaicity being transferred on a substrate of better surface quality from a worse substrate.

Mosaic spread of the redeposited films was similar to the mosaicity of the films directly stacked on glass (red circles and red triangles in Figure 8). It demonstrated how robust and reliable the graphite films were. Due to enhanced flexibility, the film follows the substrate roughness and that made the optics characteristics very sensitive to the substrate quality.

A good illustration of high stability of the deposition method is the results for 30 µm thick sample given at Figure 8. The film has anomalous high mosaicity in comparison to other films of the series obviously due to a local structural defect. The deviation of drop down value was similar on both substrates. It looks like the redeposition just aligns the structural units of the film along the substrate surface. The film replicated the surface roughness without any structural changes within the film.

Contrary to mosaic spread, the integral reflectivity does not show significant dependence on substrate material (Figure 9).

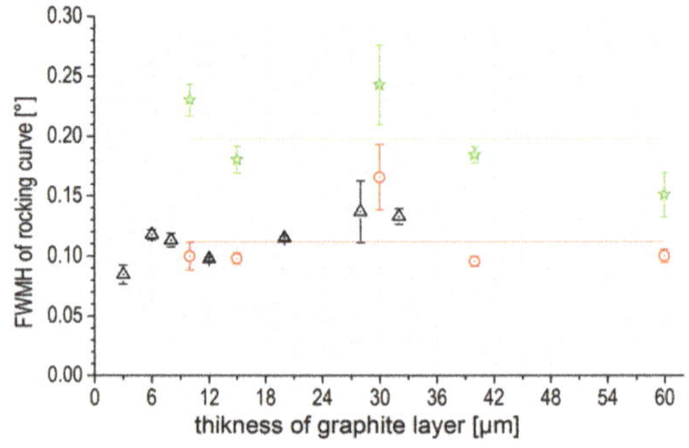

Figure 8. Mosaic spread of GrO on different substrates depending on graphite layer thickness: green stars—HAPG layer applied on aluminum; red circles—HAPG layer applied on glass (the same layer and the same measurement points as by the green stars); black triangles—HAPG layer directly applied on glass. The error bars represent the standard deviations of every 10 measurement points.

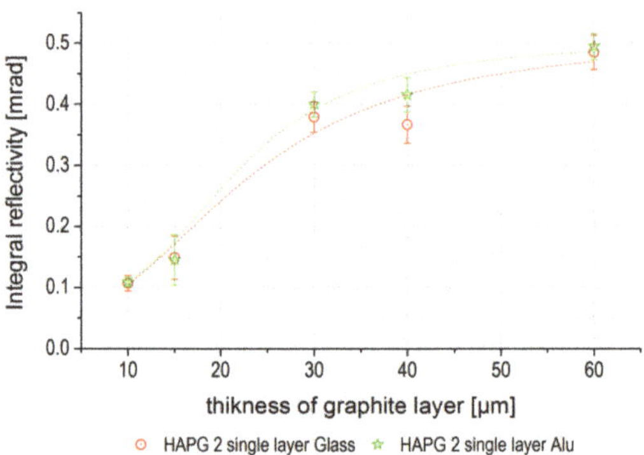

Figure 9. Integral reflectivity of GrO on different substrates depending on graphite layer thickness: green stars—HAPG layer applied on aluminum; red circles—HAPG layer applied on glass (the same layer and the same measurement points as by the green stars).

4. Materials and Methods

The measurements were fulfilled by the diffractometer D8 ADVANCE ECO from Bruker (Bruker AXS GmbH, Karlsruhe, Germany) in Bragg-Brentano configuration. As X-ray source, a tube with copper anode with 40 keV and 25 µA was used. Göbel-mirror and 2-bounce monochromator formed a parallel beam with about 1 mm × 9 mm spot in the sample plane. Measurements of the rocking curve of GrO were performed in reflection order 002 with the Cu K_a emission of the Cu anode within $\Theta = 13.29° \pm 2°$ by the SSD160 camera. To calculate the integral reflectivity, the intensity of incident X-ray beam was directly measured by 2-theta scan around 0° with SSD160.

The thin films of HOPG/HAPG with different thickness were prepared by splitting off from the same batches of bulk materials (Optigraph, Berlin, Germany). As substrates optically polished BK7-glass plates from Thorlabs GmbH Dachau/Munich, Bavatia, Germany and aluminum plates from

Kugler, Germany with surface roughness R_a = 5 nm (specified by producer) were used. Different alloys of aluminum were tested and the one (Rapidly Solidified Aluminum-6061) with the best adhesion to graphite films was used for further experiments. For direct measurement of the rocking curve, all substrates were flat.

Each measurement value depicted on the figures represents an average of single measurements of Full Width at Half Maximum (FWHM) of rocking curve made at 10 different points along the sample. In the experiment comparing the different forms of PG on glass substrate, these points were fixed relative to the substrate and not to the position on the varying graphite layers. In the experiment with glass and aluminum substrates, the points were fixed relative to their position on graphite layer, which was deposited on both substrates subsequently.

5. Conclusions

Optimization of the Graphite Optics to a given task and set-up includes not only the choice of PG and the estimation of the optics shape, but also the optimization of deposition process in order to get the combination of thickness and mosaicity of the reflecting layer required for maximizing the optics efficiency.

The correlation between the mosaicity and the thickness in HOPG optics restricts the set of existing combinations. The number of combinations could be increased by producing HAPG optics in the way similar to HOPG optics as a set of monofilms. The increase of the mosaicity fills the gap between the two types of the optics. As a result, almost any combination of thickness and mosaicity could be fulfilled and any optics according to theoretical estimation could be realized.

The quality of the substrate surface could be a crucial parameter that influences the optics performance. Therefore, the availability of the substrate of required shape and surface quality usually determines the possibility to manufacture the optics according to the requirements.

Graphite Optics is a powerful and versatile device for a wide set of applications. Due to its high variability, it could find a place in many new branches of advanced science and technology and could encourage the researchers to realize their current and future ideas.

Author Contributions: Conceptualization, project administration, funding acquisition, resources: I.G. and A.A.; methodology, data curation, investigation, validation, formal analysis, visualization: G.G.; writing—original draft preparation, writing—review and editing: I.G., A.A. and G.G.

Funding: This research received no external funding.

Conflicts of Interest: The authors declare no conflict of interest.

References

1. Malzer, W.; Grötzsch, D.; Gnewkow, R.; Schlesiger, C.; Kowalewski, F.; Van Kuiken, B.; DeBeer, S.; Kanngießer, B. A laboratory spectrometer for high throughput X-ray emission spectroscopy in catalysis research. *Rev. Sci. Instrum.* **2018**, *89*, 113111. [CrossRef]

2. Scordo, A.; Curceanu, C.; Miliucci, M.; Shi, H.; Sirghi, F.; Zmeskal, J. VOXES: A high precision X-ray spectrometer for diffused sources with HAPG crystals in the 2–20 keV range. *J. Inst.* **2018**, *13*, C04002. [CrossRef]

3. Kolmogorov, Y.; Trounova, V. Analytical potential of EDXRF using toroidal focusing systems of highly oriented pyrolytic graphite (HOPG). *X-Ray Spectrom.* **2002**, *31*, 432–436. [CrossRef]

4. Chevallier, P.; Brissaud, I.; Wang, J.X. Quantitative analysis by synchrotron radiation induced X-ray fluorescence at lure. *Nuclear Instrum. Methods Phys. Res. Sect. B Beam Interact. Mater. Atoms* **1990**, *49*, 551–554. [CrossRef]

5. Pease, D.M.; Daniel, M.; Budnick, J.I.; Rhodes, T.; Hammes, M.; Potrepka, D.M.; Sills, K.; Nelson, C.; Heald, S.M.; Brewe, D.I.; et al. Log spiral of revolution highly oriented pyrolytic graphite monochromator for fluorescence x-ray absorption edge fine structure. *Rev. Sci. Instrum.* **2000**, *71*, 3267–3273. [CrossRef]

6. Heckel, J.; Wissmann, D. Optimization of ED-XRF Excitation Configuration Parameters to Determine Trace-Element Concentrations in Organic and Inorganic Sample Matrices. *Spectroscopy* **2016**, *31*, 24–29.

7. Schlesiger, C.; Anklamm, L.; Stiel, H.; Malzer, W.; Kanngießer, B. XAFS spectroscopy by an X-ray tube based spectrometer using a novel type of HOPG mosaic crystal and optimized image processing. *J. Anal. Atom. Spectrom.* **2015**, *30*, 1080–1085. [CrossRef]

8. SIGRAY QuantumLeap-XAS, X-ray Absorption Spectroscopy System. Available online: https://www.qd-uki.co.uk/admin/images/uploaded_images/%5BQuantumLeap_Brochure% 5D20180913_QuantumLeap_Brochure-cmp.pdf (accessed on 9 January 2019).

9. Döppner, T.; Kritcher, A.L.; Neumayer, P.; Kraus, D.; Bachmann, B.; Burns, S.; Falcone, R.W.; Glenzer, S.H.; Hawreliak, J.; House, A.; et al. Qualification of a high-efficiency, gated spectrometer for X-ray Thomson scattering on the National Ignition Facility. *Rev. Sci. Instrum.* **2014**, *85*, 11D617. [CrossRef]

10. Optigraph: HOPG, Highly Oriented Pyrolytic Graphite, Graphite Monochromator, Doubly Bent Graphite Monochromator. Available online: http://www.optigraph.eu/ (accessed on 26 November 2018).

11. Je, J.H.; Lee, J.-Y. How is pyrolytic carbon formed? Transmission electron micrographs which can explain the change of its density with deposition temperature. *Carbon* **1984**, *22*, 317–319. [CrossRef]

12. Tombrel, F.; Rappeneau, F. Préparation et Structure des Pyrocarbones. In *Les Carbones; Collection de Chimie Physique*; Mason et Cie.: Paris, France, 1965; pp. 783–836. (In French)

13. Jenkins, G.M. Basal plane distortion in pyrolytic carbon. *Carbon* **1969**, *7*, 9–14. [CrossRef]

14. Moore, A.W. Highly oriented pyrolitic graphite. In *Chemistry and Physics of Carbon*; Dekker: New York, NY, USA, 1973; Volume 11, pp. 69–185. ISBN 978-0-8247-6051-9.

15. Legall, H.; Stiel, H.; Arkadiev, V.; Bjeoumikhov, A.A. High spectral resolution x-ray optics with highly oriented pyrolytic graphite. *Opt. Express* **2006**, *14*, 4570–4576. [CrossRef] [PubMed]

16. Haschke, M. *Laboratory Micro-X-Ray Fluorescence Spectroscopy: Instrumentation and Applications*; Springer Series in Surface Sciences; Springer International Publishing: Heidelberg, Germany, 2014; Volume 55, ISBN 978-3-319-04863-5.

17. Yaakobi, B.; Burek, A. Crystal diffraction systems for X-ray spectroscopy, imaging, and interferometry of laser fusion targets. *IEEE J. Quantum Electron.* **1983**, *19*, 1841–1854. [CrossRef]

18. Hoheisel, M.; Lawaczeck, R.; Pietsch, H.; Arkadiev, V. Advantages of monochromatic x-rays for imaging. In Proceedings of the Medical Imaging 2005: Physics of Medical Imaging; International Society for Optics and Photonics: San Diego, CA, USA, 2005; Volume 5745, pp. 1087–1096.

19. Jost, G.; Golfier, S.; Lawaczeck, R.; Weinmann, H.-J.; Gerlach, M.; Cibik, L.; Krumrey, M.; Fratzscher, D.; Rabe, J.; Arkadiev, V.; et al. Imaging-therapy computed tomography with quasi-monochromatic X-rays. *Eur. J. Radiol.* **2008**, *68*, S63–S68. [CrossRef] [PubMed]

20. Tuffanelli, A.; del Rio, M.S.; Pareschi, G.; Gambaccini, M.; Taibi, A.; Fantini, A.; Ohler, M. Comparative characterization of highly oriented pyrolytic graphite by means of diffraction topography. In Proceedings of the SPIE's International Symposium on Optical Science, Engineering, and Instrumentation, Denver, CO, USA, 18–23 July 1999; pp. 192–199.

21. Antonov, A.; Arkadiev, V.; Beckhoff, B.; Erko, A.; Grigorieva, I.; Kanngießer, B.; Vidal, B. 3.4 Optics for Monochromators. In *Handbook of Practical X-ray Fluorescence Analysis*; Springer: Berlin, Germany; New York, NY, USA, 2006; pp. 142–198. ISBN 978-3-540-28603-5.

22. Katrakova, D. Anwendungen der Orientierungsabbildenden Mikroskopie zur Gefügecharakterisierung kristalliner Werkstoffe. Ph.D. Thesis, Universität des Saarlands, Saarbrücken, Germany, 2002. (In German)

23. Zastrau, U.; Woldegeorgis, A.; Förster, E.; Loetzsch, R.; Marschner, H.; Uschmann, I. Characterization of strongly-bent HAPG crystals for von-Hámos x-ray spectrographs. *J. Instrum.* **2013**, *8*, P10006. [CrossRef]

24. Gerlach, M.; Anklamm, L.; Antonov, A.; Grigorieva, I.; Holfelder, I.; Kanngießer, B.; Legall, H.; Malzer, W.; Schlesiger, C.; Beckhoff, B. Characterization of HAPG mosaic crystals using synchrotron radiation. *J. Appl. Crystallogr.* **2015**, *48*, 1381–1390. [CrossRef]

25. Boslett, J.A.; Towns, R.L.R.; Megargle, R.G.; Pearson, K.H.; Furnas, T.C. Determination of parts per billion levels of electrodeposited metals by energy dispersive x-ray fluorescence spectrometry. *Anal. Chem.* **1977**, *49*, 1734–1737. [CrossRef]

26. Dresselhaus, M.S.; Dresselhaus, G. Intercalation compounds of graphite. *Adv. Phys.* **2002**, *51*, 1–186. [CrossRef]

27. Freund, A.K. Mosaic crystal monochromators for synchrotron radiation instrumentation. *Nuclear Instrum. Methods Phys. Res. Sect. A Accel. Spectrom. Detect.Assoc. Equip.* **1988**, *266*, 461–466. [CrossRef]

28. Grigorieva, I.G.; Antonov, A.A. HOPG as powerful x-ray optics. *X-Ray Spectrom.* **2003**, *32*, 64–68. [CrossRef]

29. Legall, H.; Stiel, H.; Antonov, A.; Grigorieva, I.; Arkadiev, V.; Bjeoumikhov, A.; Erko, A. A New Generation of X-Ray Optics Based on Pyrolytic Graphite. In Proceedings of the FEL 2006, Berlin, Germany, 27 August–1 September 2006; pp. 798–801.
30. HORIBA MESA-7220, X-ray Fluorescence Sulfur and Chlorine. Available online: http://www.horiba.com/us/en/scientific/products/sulfur-in-oil/mesa-7220-details/mesa-7220-x-ray-fluorescence-sulfur-and-chlorine-12793/ (accessed on 16 November 2018).
31. SPECTRO XEPOS RFA-Spektrometer—EDRFA. Available online: www.spectro.de/produkte/rfa-spektrometer/xepos-spektrometer-edrfa (accessed on 19 November 2018).
32. Anklamm, L.; Schlesiger, C.; Malzer, W.; Grötzsch, D.; Neitzel, M.; Kanngießer, B. A novel von Hamos spectrometer for efficient X-ray emission spectroscopy in the laboratory. *Rev. Sci. Instrum.* **2014**, *85*, 053110. [CrossRef]
33. Ice, G.E.; Sparks, C.J. Mosaic crystal X-ray spectrometer to resolve inelastic background from anomalous scattering experiments. *Nuclear Instrum. Methods Phys. Res. Sect. A Accel. Spectrom. Detect. Assoc. Equip.* **1990**, *291*, 110–116. [CrossRef]
34. Zastrau, U.; Brown, C.R.D.; Döppner, T.; Glenzer, S.H.; Gregori, G.; Lee, H.J.; Marschner, H.; Toleikis, S.; Wehrhan, O.; Förster, E. Focal aberrations of large-aperture HOPG von-Hàmos x-ray spectrometers. *J. Instrum.* **2012**, *7*, P09015. [CrossRef]
35. Ao, T.; Harding, E.C.; Bailey, J.E.; Loisel, G.; Patel, S.; Sinars, D.B.; Mix, L.P.; Wenger, D.F. Relative x-ray collection efficiency, spatial resolution, and spectral resolution of spherically-bent quartz, mica, germanium, and pyrolytic graphite crystals. *J. Quant. Spectrosc. Radiat. Transf.* **2014**, *144*, 92–107. [CrossRef]

Article

DAFNE-Light DXR1 Soft X-Ray Synchrotron Radiation Beamline: Characteristics and XAFS Applications

Antonella Balerna

INFN, Laboratori Nazionali di Frascati, 00044 Frascati (RM), Italy; antonella.balerna@lnf.infn.it;
Tel.: +39-06-94032542

Received: 21 November 2018; Accepted: 3 January 2019; Published: 8 January 2019

Abstract: X-ray Absorption Fine Structure Spectroscopy (XAFS) is a powerful technique to investigate the local atomic geometry and the chemical state of atoms in different types of materials, especially if lacking a long-range order, such as nanomaterials, liquids, amorphous and highly disordered systems, and polymers containing metallic atoms. The INFN-LNF DAΦNE-Light DXR1 beam line is mainly dedicated to soft X-ray absorption spectroscopy; it collects the radiation of a wiggler insertion device and covers the energy range from 0.9 to 3.0 keV or the range going from the K-edge of Na through to the K-edge of Cl. The characteristics of the beamline are reported here together with the XAFS spectra of reference compounds, in order to show some of the information achievable with this X-ray spectroscopy. Additionally, some examples of XAFS spectroscopy applications are also reported.

Keywords: soft X-rays; XAFS; beamlines; synchrotron radiation; material science

---

1. Introduction

Soft X-rays ranging from 0.9 to 3.0 keV cover an important energy range because they can be used in the study of materials containing atoms like magnesium, aluminum, silicon, sulfur, and many others. These atomic elements have an important role in fields like biology, medicine, catalysis, cultural heritage, materials, and space science. Furthermore, soft X-ray beamlines can be used for the characterization of samples using X-ray spectroscopies, but also for tests of optics and detectors needed for soft X-ray applications in other fields like space science.

X-ray Absorption Fine Structure (XAFS) spectroscopy is particularly useful to investigate the electronic structure and local environment of specific atoms in quite different samples like solids, liquids, and gasses. XAFS spectra come from the X-ray induced transition of electrons from inner-shell orbitals to unoccupied electronic states, and from the scattering of the photo-emitted core electrons by all the neighboring atoms. For samples containing light elements, in the soft X-ray region, K absorption edges can be studied, due to the excitation of 1s electrons; however for the ones containing heavier elements like Mo, Au, and so on, L or M absorption edges can be used to achieve important information on their valence band structures.

The DXR1 soft X-ray beamline is one of the beamlines of the DAΦNE-Light [1] Instituto Nazionale di Fisica Nucleare (INFN) Laboratori Nazionali di Frascati (LNF) synchrotron radiation facility. The Double Annular Φ-factory for Nice Experiments (DAΦNE) [2] storage ring is a high-luminosity, 0.51 GeV, e$^+$-e$^-$ collider, designed for a broad high-energy physics program. Due to its low-energy and high-electron current (higher than 1.5 A), DAΦNE provides high-flux synchrotron radiation (SR) beams in the energy range from IR to soft X-rays, and for this reason it is being used, in both dedicated and parasitic mode, as well as for SR applications.

In this paper, the characteristics of the DXR1 beamline, together with some XAFS measurements to show its performance, will be reported.

2. The DXR1 Beamline

The DAΦNE DXR1 soft X-ray beamline, mainly dedicated to X-ray absorption spectroscopy, started delivering beamtime to users at the end of 2004. The radiation source of the DXR1 beamline is one of the four planar wigglers (6-poles equivalent) installed on the DAΦNE electron storage ring to control the beam emittance [3]. The wiggler forces the accelerated electrons to emit a wide, intense, and polarized fan of electromagnetic radiation. Due to the wiggler higher magnetic field, the critical energy of the emitted synchrotron radiation spectrum of the DXR1 beamline (296 eV) is higher than the one of the bending magnet beamlines (219 eV). The 6 poles of the wiggler and the high storage ring current >1.5 A of DAΦNE, give a useful X-ray flux (Figure 1) for measurements well beyond ten times the critical energy.

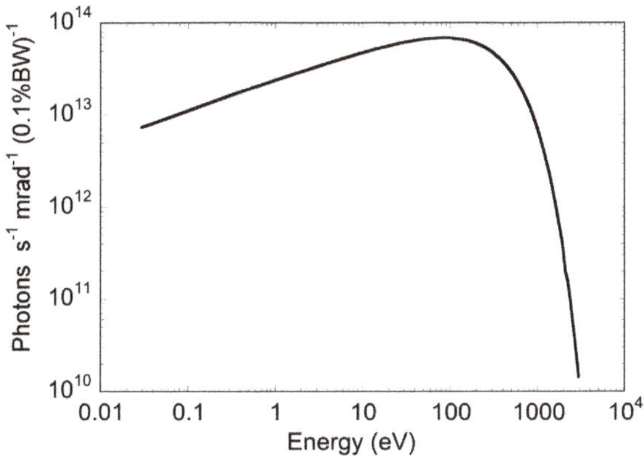

Figure 1. Calculated photon flux of the DXR1 DAΦNE wiggler (log scale), taking into account an electron circulating current of 1 A.

2.1. The Beamline Layout

A schematic view of the soft X-ray beamline is shown in Figure 2. The front end of the beamline is placed at about 4 m from the wiggler and its optical axis is geometrically aligned to the insertion device. The exit flange was designed to accept the entire vertical SR divergence (1 mrad), and about 12 mrad in the horizontal plane.

A gold-coated silicon mirror, at a grazing angle of about 2.2 degrees, deflects, in the horizontal plane, half of the beam into the UV-VIS DXR2 branch line. A removable thin high-transmittance window (8 µm Be) separates the Ultra High Vacuum (UHV) of the machine from the HV of the rest of the beamline. A double wire beam monitor can be used to control the beam position. To define the beam shape and dimensions, remotely controlled vertical and horizontal slits were installed before and after the soft X-ray monochromator, very near to the experimental chamber. The beam size used clearly depends on the dimensions of the samples to be measured; a standard one is about 2 mm in the vertical length and 8 mm in horizontal length.

To select the soft X-ray energies, the beamline is equipped with a Toyama double-crystal monochromator (Figure 3) in 'boomerang' geometry, that ensures a fixed beam exit at all achievable energies and can cover Bragg angles from 15° to 75°. The monochromator is at about 30 m from the exit flange of the front end. Different sets of crystals (see Table 1) can be used to cover the available photon energy range (0.9–3 keV). To change the crystals, the UHV chamber of the monochromator must be opened and the operation normally takes several hours.

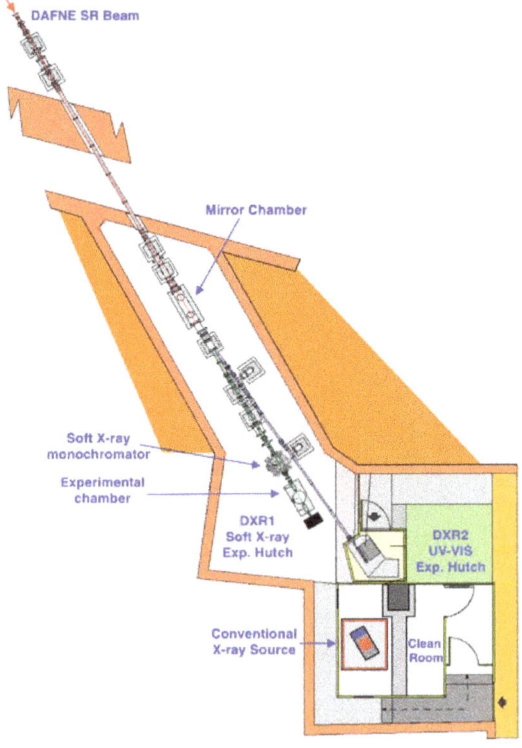

Figure 2. Schematic view of the DXR1 Soft X-ray and DXR2 UV-VIS beamlines.

Figure 3. DXR1 Soft X-ray beamline monochromator and experimental chamber from left to right.

Table 1. Available sets of X-ray crystals.

Crystal	2d Spacing (Å)	Energy Range (eV)	Absorption Edges
Beryl (10-10)	15.954	900–1560	Na K, Mg K
KTP (011)	10.950	1200–2200	Al K
InSb (111)	7.481	1800–3000	Si K–Cl K
Ge (111)	6.532	2100–3000	P K–Cl K

The typical energy resolution (E/ΔE) of the monochromator is about 1500 for Beryl and InSb, while the flux measured at 1300 eV using the Beryl crystal is 6×10^8 ph/s and 3×10^8 ph/s at 2500 eV using the InSb crystals. In both cases, the first ionization chamber was filled with N_2 gas, the gas pressure was chosen to have 10% efficiency, and the beam dimensions were (2×8) mm^2.

2.2. The Beamline Experimental Setup

A multipurpose experimental HV chamber (Figure 3), placed at about 1 m from the monochromator, was realized to allocate several samples to be measured in transmission, fluorescence, and total electron yield mode. At the moment, the only allowed mode is the transmission one where the incoming and outgoing X-ray beams are monitored using two ionization chambers. The sample holder normally used at RT can host up to ten samples. The experimental chamber can also host other kinds of sample holders, but only with a maximum dimension of about 10×10 cm^2.

From 2019 onwards, it will also be possible to perform XAFS measurements in fluorescence mode. A new 4-channel array of Silicon Drift Detectors (SDDs), called ARDESIA and developed by INFN and the Politecnico di Milano [4], has been tested on the beamline in February 2018 and will be definitively installed by the end of the year. The ARDESIA detector, having a finger-like structure, can be introduced in the experimental chamber using a specific vacuum-tight translating system and uses as an entrance window an AP5 MOXTEK thin polymer with high transmission in the soft X-ray region. In Figure 4, the first XAFS spectrum of a Pyrex thin glass, taken in fluorescence mode at the Si K-edge, is shown. The values reported for the fluorescence mode are the average values of the data measured by the four ARDESIA SDD detectors and are compared with the data taken on a Pyrex powder sample measured in transmission mode. This new detector will open the possibility to also accept experimental proposals on diluted and supported samples.

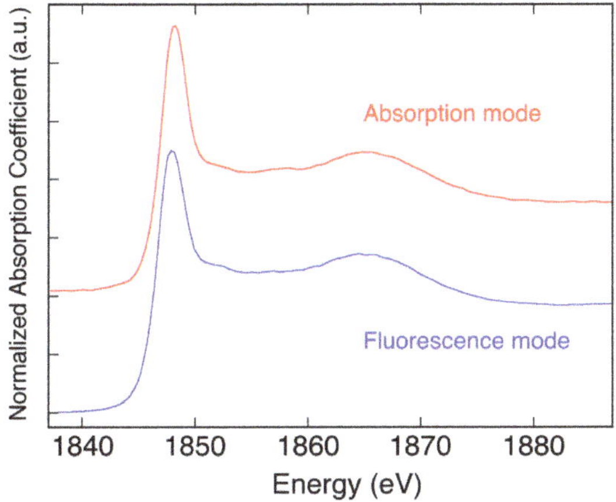

Figure 4. Comparison between the XAFS spectra of Pyrex samples measured at the DXR1 beamline in absorption and fluorescence mode.

In order to control the sample temperature and open the beamline to experimental proposals requiring low temperatures, one can use an OXFORD Instruments cryostat, which can work from 4 K up to room temperature, giving the possibility to perform XAFS measurements as a function of temperature. The sample holder for measurements from RT to 77 K can host six samples, while the one for lower temperatures can only host three.

At the end of the DXR1 experimental hutch, there is a small hutch where a tungsten micro-focus conventional X-ray tube has been installed together with an experimental chamber; this can be used to make tests on samples or devices connected to optical or detection systems.

3. XAFS Spectroscopy and Measurements

3.1. XAFS Spectroscopy

XAFS [5,6] can help understand the physical properties of materials, giving information on their local structure. XAFS is element-selective because choosing the energy of the X-rays means choosing the atomic number, Z, of the atom whose surroundings have to be characterized. XAFS is considered core level spectroscopy, because the X-ray energies used are the ones of the deep-core electrons and not of the valence ones. As a function of energy, XAFS measures the modulations of the X-ray absorption coefficient, near and above the core-level binding energies of a specific atom (Figure 5). XAFS spectra are sensitive to the oxidation state, coordination chemistry, and to the distances, coordination numbers, and species of the atoms surrounding the selected atomic element. XAFS can be used to study ordered and disordered systems, even if very diluted. XAFS can help measure 2D interatomic distances with high resolution, but also has 3D structural sensitivity. XAFS spectroscopy can be applied in the study of nanostructures, thin films, interfaces, alloys, dopants, liquids and many other very important fields, such as life-science, catalysis, cultural heritage, material, and space science.

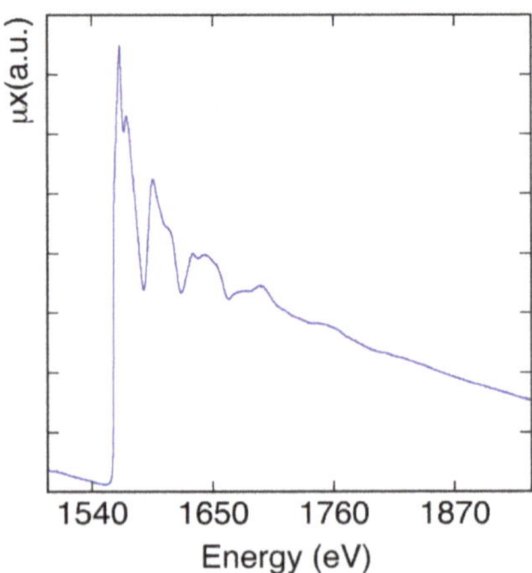

Figure 5. XAFS absorption spectrum at the K-edge of aluminum metal measured at the DXR1 beamline at room temperature.

When an X-ray beam passes through a sample, normally its intensity decreases by an amount related to the absorption characteristics of the sample itself, especially for photon energies between 1 keV and 50 keV, mainly used in XAFS spectroscopy. The mechanism contributing to the X-ray attenuation is the photoelectric absorption, resulting in the absorption of photons and emission of photoelectrons. The intensity of the transmitted X-ray beam (I_1) is related to the intensity of the incoming beam (I_0) by the Beer's Law:

$$I_1 = I_0 \exp[-\mu(E)x], \tag{1}$$

where $\mu(E)$ is the linear absorption coefficient as a function of energy, and x is the thickness of the sample; a typical XAFS setup is shown in Figure 6. I_0 and I_1 are the signals measured by the ion chambers positioned before (incoming flux) and after the sample (transmitted flux) working in transmission mode ($\mu(E)x = \ln(I_0/I_1)$. In the presence of supported or very diluted/thin samples, the transmission mode cannot be used. In this case, a fluorescence detector must be used to measure the fluorescence flux I_F ($\mu(E)x = I_F/I_0$) [5].

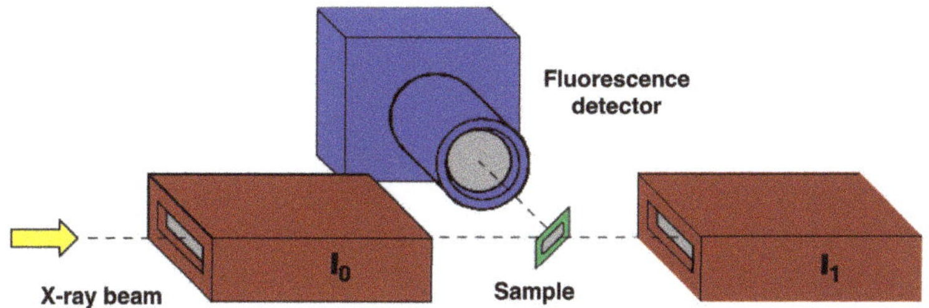

Figure 6. Schematic view of a typical XAFS experimental setup, where in transmission mode, I_0 and I_T are measured using two ionization chambers, while in fluorescence mode, a fluorescence detector needs to be used.

As clearly visible in Figure 5, where the $\mu(E)x$, evaluated using Equation (1), is reported in the low energy side of the spectrum, as the X-ray energies increase, the absorption coefficient decreases. This behavior changes at the absorption edge when the energy value of the incoming X-rays becomes enough to extract electrons from a deeper level. As shown in Figure 5, a fine structure (XAFS) starts appearing at the edge and is also present well above it. In XAFS spectra, three different regions [5] can be evidenced: the pre-edge and edge region, the near edge region or XANES (X-ray Absorption Near Edge Structure) up to about 50 eV (information on the local electronic and geometric 3D structure), and the extended region or EXAFS (Extended X-ray Absorption Fine Structure) [6], that can reach thousands of eV above the edge and can give information on the local geometric structure surrounding the absorbing atoms.

3.2. XANES and EXAFS Spectra

Even if XANES modeling [5] is very complex, important information like the oxidation state, three dimensional geometry, and coordination environment of elements under investigation can be achieved by also comparing the measured spectra with those of well-known model compounds. In Figure 7, the normalized XANES spectra measured on crystalline samples with known crystal structures, containing sulfur at different valence states, are reported.

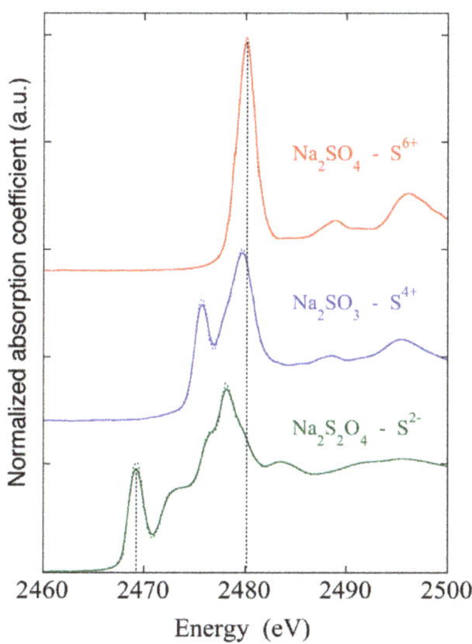

Figure 7. Energy shifts of the S K-edge as a function of S valence changes.

The shape of the edge and the pre-edge resonances are characteristic of the local symmetry of the investigated atom sites and can be used as fingerprints in the identification of the local structure of unknown samples. The binding energies of the valence orbitals and therefore the energy position of the sulfur edge are correlated with the valence state of the absorbing atom. As the oxidation state increases, the absorption edges in the XANES spectra move to higher energies. Energy shifts vary linearly with the valence of the absorbing atom [5], and in particular, as a function of the sulfur oxidation state, large energy differences up to 12 eV can be found between S^{2-} and S^{6+} [7].

XANES spectra can give chemical and structural information and can be very important in many different fields. In the field of cultural heritage, X-ray elemental micro mapping can give information on the atomic elements present in paintings; but when using XAFS spectroscopy, it becomes also possible to achieve information on the chemical composition in the presence of trace elements as well [8]. Due to the very different features present in the XANES spectra of elements in metallic or different oxide phases (see Figure 8), sometimes XANES spectra can directly give the required information on the chemical state of the materials being studied [9].

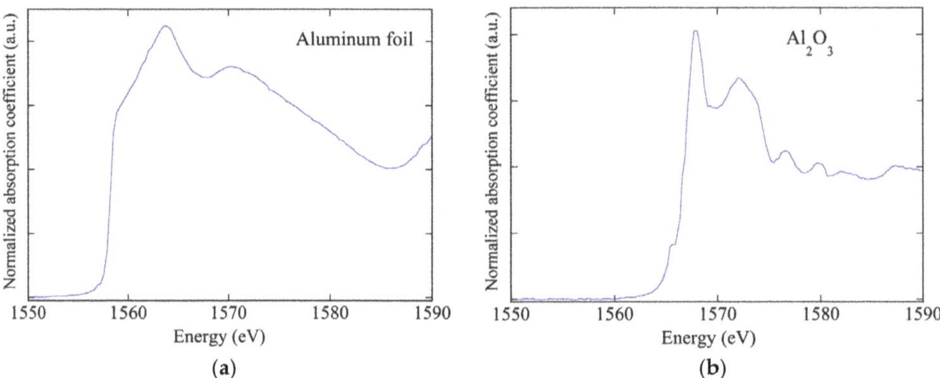

Figure 8. The very different XANES spectra of: (**a**) aluminum metal foil and (**b**) an aluminum oxide Al$_2$O$_3$.

From the analysis of the EXAFS data, quantitative information on the short-range order structure can be achieved. EXAFS oscillations are usually indicated as $\chi(k)$:

$$\chi(k) = \frac{\mu(k) - \mu_0(k)}{\mu_0(k)}, \tag{2}$$

where $\mu_0(k)$ is the monotonically decreasing atomic absorption coefficient, $\mu(k)$ is the effective absorption coefficient oscillating around it, and k is the photoelectron wave vector [10] given by Equation (3)

$$k = \sqrt{\frac{2m(E - E_0)}{\hbar^2}}, \tag{3}$$

where E is the incoming photon energy, and E_0 is the absorption edge energy calculated as the energy of the maximum derivative of $\mu(E)$. EXAFS oscillations can be well approximated by Equation (4):

$$\chi(k) = \sum_j \frac{S_0^2 N_j f_j(k) e^{-2k^2 \sigma_j^2}}{k R_j^2} \sin(2k R_j + \delta_j(k)). \tag{4}$$

In $\chi(k)$, the backscattering atoms around the absorbing one are grouped in coordination shells [5,6], each one containing a number, N_j of atoms, of the same species, at the same distance R_j from the absorbing atom. $\chi(k)$ is formally given by the sum over the index j of the contributions coming from the j coordination shells [5,6]. In Equation (4), $f_j(k)$ and $d_j(k)$ are scattering properties of the atoms around the absorbing one and in particular $f_j(k)$ is the backscattering amplitude of the N_j neighbors, while $d_j(k)$ is a k-dependent phase shift [5,6]. As shown in Figure 9, EXAFS data can be extracted using programs like the ATHENA program [11], after the linear subtraction of the pre-edge background and the removal of the atomic absorption.

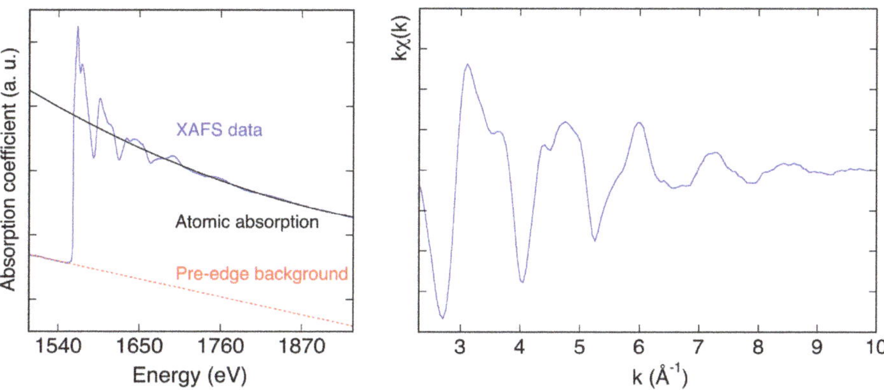

Figure 9. **Left side**: Data extraction process for the analysis of the EXAFS data of an aluminum metal foil pre-edge background subtraction and normalization; **Right side**: Extracted EXAFS data of an aluminum metal foil.

Since the EXAFS signal contains the contributions of all of the j coordination shells and each contribution can be approximated by a damped sinusoidal function in k-space whose frequency is proportional to a specific absorber-backscatterer distance, Fourier transforming (FT) the EXAFS data enables the separation of the different frequencies. This operation transforms each EXAFS sinusoidal component in a FT modulus peak, going from the $k(\text{Å}^{-1})$ space to an R(Å) space. The height of the peaks depends on the amplitude parameters of the EXAFS equation, while their position depends on the phase parameters. In Figure 10 the FT of the EXAFS spectrum of an Aluminum foil is reported: the first and second peaks represent the first and second coordination shells. Having aluminum a face centered cubic (fcc) structure [12], the coordination number of the nearest neighbors is 12, while the second shell coordination number is 6.

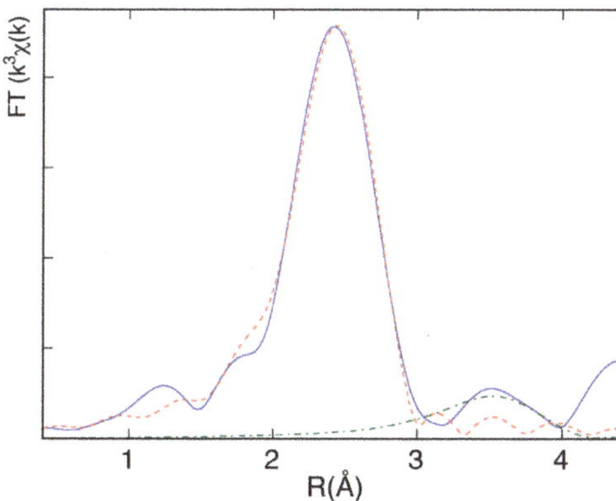

Figure 10. Fourier Transforms of the EXAFS experimental data (full line) and of the theoretical first (red dashed lines) and second (green dashed lines) shell contributions.

A fitting procedure can be used to determine the coordination numbers (N_j), the interatomic distances (R_j) and Debye–Waller thermal and static disorder factors (σ^2) of the coordination shells around the absorbing atom.

The least-square fitting of the structural parameters can be performed using a program like ARTEMIS [11] that—together with the ATHENA program—is implemented in the IFEFFIT package [13]. In the fitting procedures, the scattering contributions can be calculated by software packages like FEFF [14,15], and depend on the coordination shells around the absorbing atom that must be studied. In all cases, an estimation of the accuracy of the obtained structural parameters, compatible with data quality and range used [16] is also normally evaluated. In Table 2, the results achieved in the fitting procedure of the first two Al coordination shells are reported. The results achieved by the XAFS data taken at room temperature, are in good agreement with the fcc Al values [17].

Table 2. Results achieved in the fitting procedure of the first two Al coordination shells.

Shell	N	R (Å)	σ^2 (Å^2)
First	12	2.85(1)	0.014(3)
Second	6	4.04(1)	0.027(4)

3.3. Using the Beamline for XAFS Applications

Twice a year, a call for proposals is open to EU, Italian, and other external users coming from Universities or Research Centers. The transnational access to all the DAFNE-Light beamlines open to users is nowadays supported by the EU CALIPSOplus project [18]. In recent years, the soft X-ray line has successfully delivered beamtime to many different experimental proposals. The X-ray Absorption Near Edge Spectroscopy (XANES) technique has been routinely applied in transmission mode on different samples within the energy range 1.0–3.0 keV. XANES spectra were acquired in parasitic and in dedicated beamtime days. During the parasitic mode days, tests of new samples and experiments not requiring long acquisition times, like studies of diamond detectors, soft X-ray multi-layers, and imaging of metal impurities in leaves were performed. Dedicated beamtime was normally used for selected experimental proposals chosen by the INFN-LNF DAFNE-Light User Selection Panel.

In the last year, some interesting experiments on silicate and lapis lazuli pigments at the Si and S K-edges that can have applications in the cultural heritage field have been performed and XANES data are being analyzed.

The possibility to perform measurements at low temperatures was used in experiments requiring tests of systems needed for space applications. In particular, the thermal characterization of the X-ray transmission of thin aluminum filters, needed to protect X-ray detectors for space missions, was performed at the DXR1 beamline. These measurements are important to characterize the effects induced on the detector by the aluminum filters. As shown in Figure 11 (left panel), where the EXAFS spectra of an aluminum foil as a function of temperature are reported, X-ray transmission measures the presence of a fine structure, but also of thermal effects that affect the phase and amplitude of the EXAFS oscillations. Both effects are probably more evident in Figure 11 (right panel), where the Fourier transforms of these spectra, calculated in the k range (2–7) Å^{-1}, are reported. The reduction of the intensity of the peaks corresponding to the different coordination shells and the shift of their positions in R space, as a function of temperature, are now clearly more visible.

Just to give an idea of other kind of measurements that can be performed at the DXR1 beamline, some interesting applications will be reported concerning hydrogen storage materials [19], thiol-capped gold nanoparticles [20], and anticancer metallodrugs [21].

Concerning hydrogen storage materials, tetrahydroaluminates or alanates, complex hydrides containing AlH$_4$ groups, were studied [19] for the development of higher-efficiency hydrogen storage materials, since it was discovered that the hydrogen de- and absorption can be catalyzed by doping with titanium and other transition metal and rare earth metal species. The aim of the experiments at the Al K edge was to investigate the local structure around Al atoms studying the alanate phase at different stages of the reaction after the material has been cycled under hydrogen. The study allowed showing the presence of modification within the alanate structure during cycling under hydrogen.

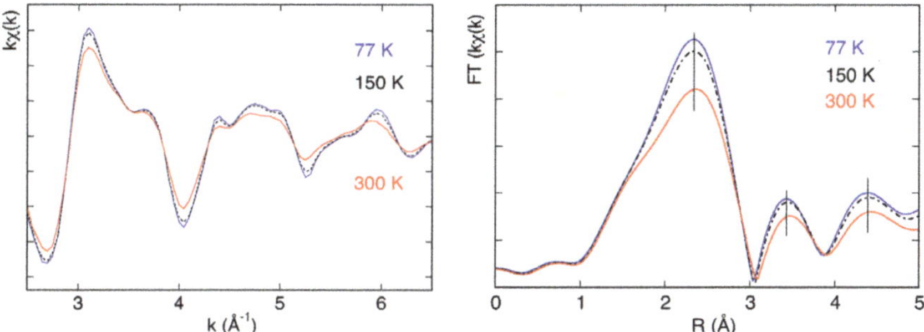

Figure 11. EXAFS spectra (**left panel**) of an aluminum foil measured at 77 K, 150 K and 300 K) and their Fourier transforms (**right panel**).

Moving to nanomaterials, it is well-known that the study on their size dependent structural and electronic properties asks, as ideal condition, nearly monodisperse particles. One way to achieve this is by capping nanoparticles with molecular species that interact with the surface preventing the nucleation or aggregation of single clusters. Capping molecules generally contain functional groups such as amine, alcohol, thiol, and phosphine, providing a wide range of interactions. Thiol-capping molecules are particularly suitable in preparative methods based on the chemical synthesis of nanoparticles, such as SMAD [20]. In general, thiols interact strongly with a gold surface, inducing meaningful charge redistribution. A thiol–Au interaction is quite important because it exhibits the interesting property of self-assembly. In principle, alkanethiolates are dissociatively chemisorbed to a gold surface via the sulfur atom after cleavage of S–H bonds. When interacting with a gold surface, different structural phases occur at increasing coverage; in particular, high coverage results in a formation of ordered structures. To better understand the gold–sulfur interactions, the XAFS sulfur K-edge measurements were performed at the DXR1 beamline (Figure 12), while the Au L_3 measurements were performed elsewhere.

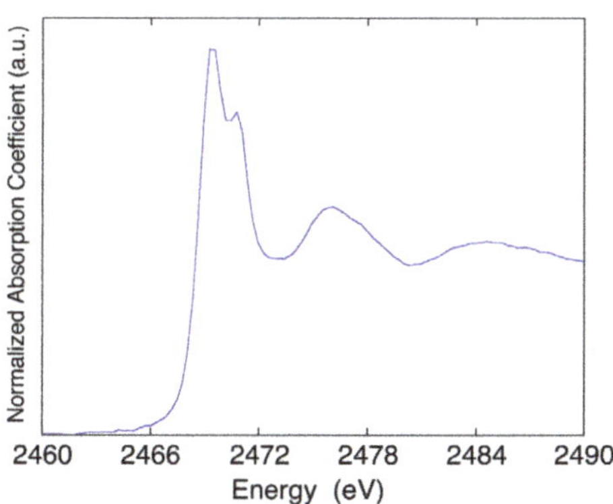

Figure 12. XANES spectrum at S K-edge of a gold thiol-capped sample.

Moving to a totally different field, an important study has been performed on adducts of ruthenium anticancer metallodrugs with serum proteins and fragments of proteins [21]. There is a great interest in the analysis of the interactions of metal-based drugs with serum proteins in view of their relevant biological and pharmacological implications. Specifically, great attention has been given to ruthenium complexes that seem to be very promising.

The mechanisms through which the metal complexes produce their biological and pharmacological effects are still largely unexplored and it seems that ruthenium complexes act on different targets, most likely on proteins.

The reaction of bovine serum albumin (BSA) with [*trans*-RuCl$_4$ (Im)(dimethylsulfoxide)][ImH] (Im = imidazole) (NAMI-A), an experimental ruthenium(III) anticancer drug, and the formation of the respective NAMI-A/BSA adduct have been investigated at the DXR1 beamline by XAFS measurements at the sulfur and chlorine K-edges and at the ruthenium L$_3$ edge. Ruthenium XAFS data proved unambiguously that the ruthenium remains in the oxidation state, Ru(III), after protein binding. Comparative analysis of the chlorine K-edge XAS spectra of NAMI-A and NAMI-A/BSA revealed that the chlorine environment was greatly perturbed upon protein binding (Figure 13). Only small changes were observed in the sulfur K-edge spectra (Figure 13), probably because it was dominated by several protein sulfur groups. Valuable information on the nature of this metallodrug/protein adduct and on the mechanism of its formation was gained, and XAFS spectroscopy turned out to be a very suitable method for the study of this kind of systems.

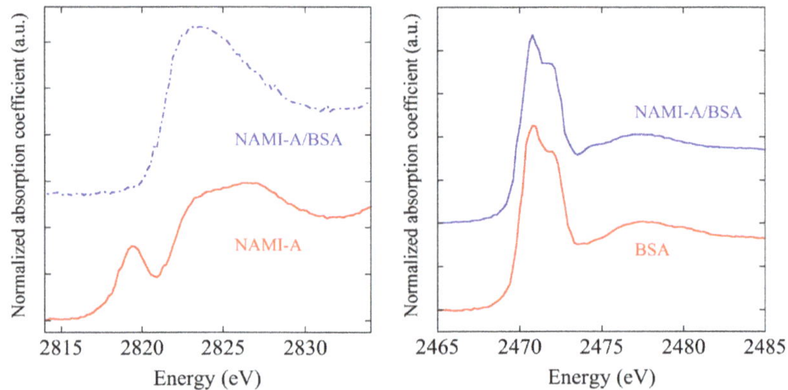

Figure 13. Left panel: normalized spectra at the chlorine K-edge of NAMI-A and of the NAMI-A/BSA adduct; **Right panel**: normalized spectra at the sulfur K-edge of bovine serum albumin (BSA) and of the NAMI-A/BSA adduct.

4. Conclusions

The INFN-LNF soft X-ray DXR1 beamline started delivering beamtime to users at the end of 2004, but during the last few years, several improvements have been made concerning the sample environment. XAFS measurements can now be performed at low temperatures and, starting from 2019, in fluorescence mode as well, opening the possibility to study diluted and supported samples. The DXR1 beamline is mainly used for XAFS spectroscopy measurements but has also been used to test detectors and optical elements. As mentioned, XAFS spectroscopy can help achieving a complete local structural characterization of different kinds of samples, being very sensitive to the formal oxidation states, coordination chemistry, distances, coordination number, and species of the atoms immediately surrounding the selected atomic elements. At the DXR1 soft X-ray beamline, XAFS spectroscopy has been applied in many different fields, some examples of which have been reported here.

Funding: This research received no external funding.

Acknowledgments: The author wants to acknowledge the technical support given by A. Grilli, M. Pietropaoli, A. Raco, G. Viviani, V. Tullio, and V. Sciarra.

Conflicts of Interest: The author declares no conflict of interest.

References

1. Balerna, A. DAΦNE-Light Facility Update. *Synchrotron Radiat. News* **2014**, *27*, 21–24. [CrossRef]
2. Zobov, M.; Alesini, D.; Biagini, M.E.; Biscari, C.; Bocci, A.; Boni, R.; Boscolo, M.; Bossi, F.; Buonomo, B.; Clozza, A.; et al. Test of "Crab-Waist" Collisions at the DA Φ NE Φ Factory. *Phys. Rev. Lett.* **2010**, *104*, 174801. [CrossRef] [PubMed]
3. Titkova, I.; Zobov, M.; Dabagov, S. Synchrotron Radiation from DAFNE Bending Magnet and Wiggler. Available online: https://www.openaccessrepository.it/record/20738/files/LNF-03-2(IR).pdf (accessed on 7 January 2019).
4. Bellotti, G.; Butt, A.D.; Carminati, M.; Fiorini, C.; Bombelli, L.; Borghi, G.; Piemonte, C.; Zorzi, N.; Balerna, A. The ARDESIA Detection Module: A 4-Channel Array of SDDs for Mcps X-ray Spectroscopy in Synchrotron Radiation Applications. *IEEE Trans. Nucl. Sci.* **2018**, *65*, 1355–1364. [CrossRef]
5. Koningsberger, D.C.; Prins, R. *X-ray Absorption: Principles and Application Techniques of EXAFS, SEXAFS and XANES*; John Wiley & Sons: New York, NY, USA, 1988.
6. Rehr, J.J.; Albers, R.C. Theoretical approaches to X-ray absorption fine structure. *Rev. Mod. Phys.* **2000**, *72*, 621–654. [CrossRef]
7. Jalilehvand, F. Sulfur: Not a "silent" element any more. *Chem. Soc. Rev.* **2006**, *35*, 1256–1268. [CrossRef] [PubMed]
8. Cotte, M.; Susini, J.; Dik, J.; Janssens, K. Synchrotron-Based X-ray Absorption Spectroscopy for Art Conservation: Looking Back and Looking Forward. *Acc. Chem. Res.* **2010**, *43*, 705–714. [CrossRef] [PubMed]
9. Li, D.; Bancroft, G.M.; Fleet, M.E.; Feng, X.H.; Pan, Y. Al K-edge XANES spectra of aluminosilicate minerals. *Am. Mineral.* **1995**, *80*, 432–440. [CrossRef]
10. Lee, P.A.; Citrin, P.H.; Eisenberger, P.; Kincaid, B.M. Extended X-ray absorption fine structure and limitations as a structural tool. *Rev. Mod. Phys.* **1981**, *53*, 769–806. [CrossRef]
11. Ravel, B.; Newville, M.; Athena, A. Hephaestus: Data analysis for X-ray absorption spectroscopy using IFEFFIT. *J. Synchrotron Radiat.* **2005**, *12*, 537–541. [CrossRef] [PubMed]
12. Jouve, A.; Nagy, G.; Somodi, F.; Tiozzo, C.; Villa, A.; Balerna, A.; Beck, A.; Evangelisti, C.; Prati, L. Gold-silver catalysts: Effect of catalyst structure on the selectivity of glycerol oxidation. *J. Catal.* **2018**, *368*, 324–335. [CrossRef]
13. Newville, M. IFEFFIT: Interactive XAFS analysis and FEFF fitting. *J. Synchrotron Radiat.* **2001**, *8*, 322–324. [CrossRef] [PubMed]
14. Zabinsky, S.I.; Rehr, J.J.; Ankudinov, A.; Albers, R.C.; Eller, M.J. Multiple-scattering calculations of X-ray-absorption spectra. *Phys. Rev. B* **1995**, *52*, 2995–3009. [CrossRef]
15. Rehr, J.J.; Kas, J.J.; Vila, F.D.; Prange, M.P.; Jorissen, K. Parameter-free calculations of X-ray spectra with FEFF9. *Phys. Chem. Chem. Phys.* **2010**, *12*, 5503–5513. [CrossRef] [PubMed]
16. Li, G.G.; Bridges, F.; Booth, C.H. X-ray-absorption fine-structure standards: A comparison of experiment and theory. *Phys. Rev. B* **1995**, *52*, 6332–6348. [CrossRef]
17. Zeng, L.; Tran, D.T.; Tai, C.-W.; Svensson, G.; Olsson, E. Atomic structure and oxygen deficiency of the ultrathin aluminium oxide barrier in Al/AlOx/Al Josephson junctions. *Sci. Rep.* **2016**, *6*, 29679. [CrossRef] [PubMed]
18. CALIPSOplus Is an H2020 Integrating Activity for Advanced Communities-Grant Agreement N. Available online: http://www.calipsoplus.eu/ (accessed on 7 January 2019).
19. Leon, A.; Balerna, A.; Cinque, G.; Frommen, C.; Fichtner, M. Al K-edge XANES Measurements in NaAlH$_4$ Doped with TiCl$_3$ by Ball Milling. *J. Phys. Chem. C* **2007**, *111*, 3795. [CrossRef]
20. Comaschi, T.; Balerna, A.; Mobilio, S. Temperature dependence of the structural parameters of gold nanoparticles investigated with EXAFS. *Phys. Rev. B* **2008**, *77*, 075432. [CrossRef]
21. Ascone, I.; Messori, L.; Casini, A.; Gabbiani, C.; Balerna, A.; Dell'Unto, F.; Congiu Castellano, A. Exploiting Soft and Hard X-Ray Absorption Spectroscopy to Characterize Metallodrug/Protein Interactions: The Binding of [trans-RuCl4(Im)(dimethylsulfoxide)][ImH] (Im = imidazole) to Bovine Serum Albumin. *Inorg. Chem.* **2008**, *47*, 8629–8634. [CrossRef] [PubMed]

Article

The Potential of EuPRAXIA@SPARC_LAB for Radiation Based Techniques

Antonella Balerna [1], Samanta Bartocci [2], Giovanni Batignani [3], Alessandro Cianchi [4,5], Enrica Chiadroni [1], Marcello Coreno [1,6], Antonio Cricenti [6], Sultan Dabagov [1,7,8], Andrea Di Cicco [9], Massimo Faiferri [2], Carino Ferrante [3,10], Massimo Ferrario [1], Giuseppe Fumero [3,11], Luca Giannessi [12,13], Roberto Gunnella [9], Juan José Leani [14], Stefano Lupi [3,15], Salvatore Macis [4,5], Rosa Manca [2], Augusto Marcelli [1,6], Claudio Masciovecchio [12], Marco Minicucci [9], Silvia Morante [4,5], Enrico Perfetto [4,16], Massimo Petrarca [3,15], Fabrizio Pusceddu [2], Javad Rezvani [1], José Ignacio Robledo [14], Giancarlo Rossi [4,5,17], Héctor Jorge Sánchez [14,18], Tullio Scopigno [3,10], Gianluca Stefanucci [4,5], Francesco Stellato [4,5,*], Angela Trapananti [9] and Fabio Villa [1]

[1] Istituto Nazionale di Fisica Nucleare INFN—Laboratori Nazionali di Frascati, via E. Fermi 40, 00044 Frascati, Italy; Antonella.Balerna@lnf.infn.it (A.B.); Enrica.Chiadroni@lnf.infn.it (E.C.); marcello.coreno@elettra.eu (M.C.); sultan.dabagov@lnf.infn.it (S.D.); Massimo.Ferrario@lnf.infn.it (M.F.); augusto.marcelli@lnf.infn.it (A.M.); Javad.Rezvani@lnf.infn.it (J.R.); fabio.Villa@lnf.infn.it (F.V.)

[2] Dipartimento di Architettura, Design e Urbanistica, Università degli studi di Sassari, Palazzo del Pou Salit—Piazza Duomo, 6-07041 Alghero, Italy; samantabartocci@gmail.com (S.B.); faiferri@uniss.it (M.F.); mancarosa@yahoo.it (R.M.); fabrizio_pusceddu@yahoo.it (F.P.)

[3] Dipartimento di Fisica, Università degli studi di Roma "La Sapienza", P.le Aldo Moro 2, 00185 Rome, Italy; giovanni.batignani@roma1.infn.it (G.B.); carino.ferrante@iit.it (C.F.); giuseppe.fumero@gmail.com (G.F.); stefano.lupi@roma1.infn.it (S.L.); massimo.petrarca@roma1.infn.it (M.P.); tullio.scopigno@uniroma1.it (T.S.)

[4] Dipartimento di Fisica, Università degli Studi di Roma Tor Vergata, Via della Ricerca Scientifica 1, 00133 Rome, Italy; Alessandro.Cianchi@roma2.infn.it (A.C.); Salvatore.Macis@roma2.infn.it (S.M.); silvia.morante@roma2.infn.it (S.M.); Enrico.Perfetto@roma2.infn.it (E.P.); Giancarlo.Rossi@roma2.infn.it (G.R.); stefanucci@roma2.infn.it (G.S.)

[5] Istituto Nazionale di Fisica Nucleare (INFN) Sezione di Roma Tor Vergata, Via della Ricerca Scientifica 1, 00133 Rome, Italy

[6] Istituto di Struttura della Materia (ISM)-Consiglio Nazionale delle Ricerche (CNR), via Fosso del Cavaliere 100, 00133 Rome, Italy; antonio.cricenti@artov.ism.cnr.it

[7] National Research Nuclear University MEPhI, Kashirskoe Sh. 31, 115409 Moscow, Russia

[8] P.N. Lebedev Physical Institute, Russian Academy of Sciences (RAS), Leninsky Pr. 53, 119991 Moscow, Russia

[9] School of Science and Technology, Physics Division, Università degli studi di Camerino, via Madonna delle Carceri, 62032 Camerino, Italy; andrea.dicicco@unicam.it (A.D.C.); roberto.gunnella@unicam.it (R.G.); marco.minicucci@unicam.it (M.M.); angela.trapananti@unicam.it (A.T.)

[10] Center for Life Nano Science @Sapienza, Istituto Italiano di Tecnologia, 00161 Rome, Italy

[11] Dipartimento di Scienze di Base e Applicate per l'Ingegneria, Università degli studi di Roma "La Sapienza" Roma, 00161 Rome, Italy

[12] Elettra-Sincrotrone Trieste, Area Science Park, 34149 Trieste, Italy; lucagiannessi@gmail.com (L.G.); claudio.masciovecchio@elettra.eu (C.M.)

[13] ENEA C.R. Frascati, Via E. Fermi 45, 00044 Frascati RM, Italy

[14] National Scientific and Technical Research Council, C1425FQB Buenos Aires, Argentina; newjuanjo@gmail.com (J.J.L.); jorobledo@unc.edu.ar (J.I.R.); jsan@famaf.unc.edu.ar (H.J.S.)

[15] Istituto Nazionale di Fisica Nucleare (INFN) Sezione di Roma La Sapienza, P.le Aldo Moro 2, 00185 Rome, Italy

[16] Istituto di Struttura della Materia (ISM)-Consiglio Nazionale delle Ricerche (CNR), Division of Ultrafast Processes in Materials (FLASHit), Area della Ricerca di Roma 1, Via Salaria Km 29.3, I-00016 Monterotondo Scalo, Italy

Condens. Matter **2019**, *4*, 30; doi:10.3390/condmat4010030

www.mdpi.com/journal/condensedmatter

[17] Centro Fermi—Museo Storico della Fisica e Centro Studi e Ricerche "Enrico Fermi", 00184 Rome, Italy
[18] Faculty of Mathematics Astronomy Physics, National University of Córdoba, Córdoba X5000GYA, Argentina
* Correspondence: francesco.stellato@roma2.infn.it

Received: 31 January 2019; Accepted: 2 March 2019; Published: 7 March 2019

Abstract: A proposal for building a Free Electron Laser, EuPRAXIA@SPARC_LAB, at the Laboratori Nazionali di Frascati, is at present under consideration. This FEL facility will provide a unique combination of a high brightness GeV-range electron beam generated in a X-band RF linac, a 0.5 PW-class laser system and the first FEL source driven by a plasma accelerator. The FEL will produce ultra-bright pulses, with up to 10^{12} photons/pulse, femtosecond timescale and wavelength down to 3 nm, which lies in the so called "water window". The experimental activity will be focused on the realization of a plasma driven short wavelength FEL able to provide high-quality photons for a user beamline. In this paper, we describe the main classes of experiments that will be performed at the facility, including coherent diffraction imaging, soft X-ray absorption spectroscopy, Raman spectroscopy, Resonant Inelastic X-ray Scattering and photofragmentation measurements. These techniques will allow studying a variety of samples, both biological and inorganic, providing information about their structure and dynamical behavior. In this context, the possibility of inducing changes in samples via pump pulses leading to the stimulation of chemical reactions or the generation of coherent excitations would tremendously benefit from pulses in the soft X-ray region. High power synchronized optical lasers and a TeraHertz radiation source will indeed be made available for THz and pump–probe experiments and a split-and-delay station will allow performing XUV-XUV pump–probe experiments.

Keywords: free electron lasers; coherent imaging; X-ray Raman; X-ray absorption; THz radiation

---

1. Introduction

The advent of Free Electron Lasers (FELs) opened up the way to an unprecedented, wide class of experiments exploiting the peculiar features of these radiation sources. Key elements are the high peak brilliance that can be higher than 10^{27} photons/(s mm^2 mrad^2 0.1% bandwidth) and the short pulse duration, which is of the order of tens of femtoseconds. FELs can therefore allow high time resolution measurements and may provide a high signal-to-noise ratio. By exploiting the high peak brilliance and the extremely short FEL pulses the so-called diffract-and-destroy regime, in which interpretable data are gathered before the sample is destroyed by the FEL pulse radiation [1], can be explored, overcoming one of the main limitations of synchrotron radiation based experiments, namely sample radiation damage. This idea has been proven in several experiments on various samples, both biological [1–6] and non-biological [7], at different wavelengths ranging from the UV to the hard X-rays region. Actually, this issue is particularly relevant since coherent diffraction imaging (CDI) of biological system using conventional methods is ultimately limited by radiation damage owing to the large amount of energy deposited in the sample by the photon beam [7,8].

The unique FEL features (energy range, time resolution and brilliance) can be exploited in several branches of physics, chemistry, material science and biology. In this paper, we describe the main experimental lines that can be investigated at the EuPRAXIA@SPARC_LAB FEL [9,10] (Figure 1). The EuPRAXIA@SPARC_LAB FEL will provide photon pulses with high intensity, up to 10^{12} photons/pulse, down to a wavelength of about 3 nm, in the so called "water window". The foreseen pulse energy will reach 180 μJ and the bandwidth will range between 0.4% and 0.9%, according to the machine operation scheme. The pulse length will be of tens of femtoseconds. The experimental activity will be focused on the realization of a plasma driven short wavelength FEL and the first expected FEL operational mode will be based on the self amplification of spontaneous radiation (SASE) mechanism with tapered undulators. Details about foreseen beam parameters are given in [9]. The facility will also provide a high-power (0.5 PW) laser system and a TeraHertz (THz) radiation source. These sources will allow performing laser pump–FEL probe and THz pump–FEL probe experiments. Moreover, a split-and-delay element will allow laser pump–FEL probe experiments. A fully equipped experimental endstation designed to perform this variety of experiments will be designed and built. The experimental hall will be designed in order to allow the highest flexibility, optimizing the available space to perform a wide class of experiments (see Figure 2). All the aspects of the experimental needs will be considered, therefore next to the experimental hall a large space for the support to the experimental activities, but also for rest breaks of people working on the experiments, will be available.

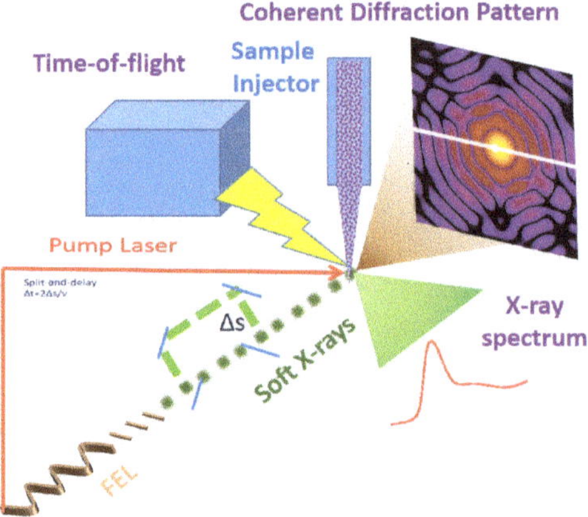

Figure 1. A simplified layout of the experiments that will be performed at EX-TRIM.

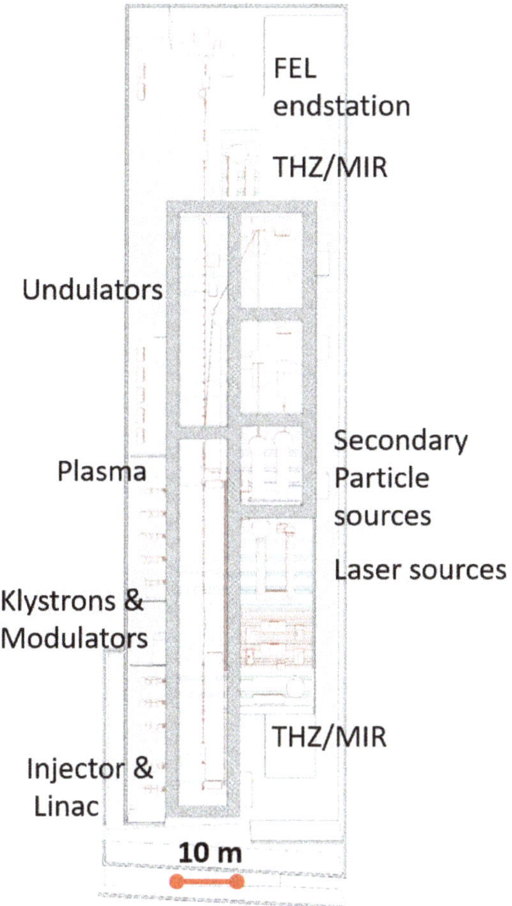

Figure 2. A layout of EuPRAXIA@SPARC_LAB. The building will be about 135 m long and 35 m wide. The location of the injector and linac, of the klystrons, of the plasma module and of undulators is shown on the left side. The location of the two THz/MIR and of the FEL endstation is highlighted on the right side.

2. Expected Results

In this section, we present and discuss some of the results that we expect to be able to obtain owing to the peculiar features of the EuPRAXIA@SPARC_LAB facility.

2.1. Coherent Imaging of Biological Samples

FEL radiation can be used to gather information on several kinds of biological samples. Biological single particle imaging is one of the main topics of FEL research with dedicated beamlines, although the FEL intensity, the available detectors, and techniques to introduce the sample into the focused X-ray sampling position, were all insufficient to obtain (near) atomic resolution structural information from single biological macromolecules [11]. Nevertheless, recent experiments performed at LCLS showed that single-shot diffraction patterns from biological samples as small as 70 nm in diameter (e.g., the enterobacteria phage PR772 [12,13]) can be measured, and still at LCLS signal up to 5.9Åresolution was observed from rice dwarf virus [14]. Measurements performed at SACLA allowed recovering the electron density of chloroplast with 70 nm resolution [15]. Dataset of coherent imaging patterns have also been made available for the scientific community [16,17]. In the specific case at hand here, thanks to the photon energy range delivered by the EuPRAXIA@SPARC_LAB FEL, coherent imaging experiments in the water window will allow obtaining structural information on cells, organelles, viruses and protein aggregates [3–5,18] by performing measurements at room temperature and with samples staying in their native state. Exploiting the high degree of transverse coherence of the EuPRAXIA@SPARC_LAB FEL beam, which is foreseen to be between 80% and 100%, 2D images of a variety of biological samples, including bacteria (see Figure 3a), viruses (see Figure 3b), cells, cell organelles and protein aggregates and fibrils [19,20], can be obtained. The possibility of obtaining high resolution structures of fibrils in native conditions is particularly relevant to study the dynamics of their formation, which is important for both industrial/pharmaceutical [21,22] and bio-medical [20] applications. When dealing with a class of identical objects (e.g., viruses or ribosomes), it is also possible to combine the diffraction patterns, coming from different FEL pulses hitting different elements of the class, to get a full 3D reconstruction [23].

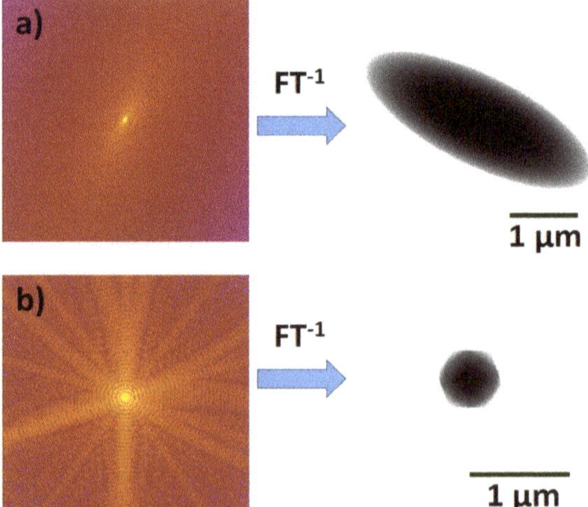

Figure 3. Simulated data for coherent imaging experiments at the EuPRAXIA@SPARC_LAB FEL. (a) A simulated diffraction pattern and the reconstruction of the electron density of a 2 µm long spheroid, with a shape similar to that of an elongated bacterium. (b) A simulated diffraction pattern from a 600 nm diameter icosahedral virus. Simulations were performed using the software Condor [24] assuming a Gaussian-shaped beam with a diameter of 3 µm, a wavelength of 2.87 nm and a pulse intensity of 100 µJ, which is, according to simulations, the expected pulse energy delivered on the sample by an EuPRAXIA@SPARC_LAB pulse.

The attainable resolution depends on the sample's composition and size and it is limited by FEL wavelength and photon brilliance, but, thanks to the high contrast associated to the water-window energy range, the EuPRAXIA@SPARC_LAB will be a suitable facility to perform CXI measurements on a wide class of biological objects. It is worth pointing out that, when dealing with biological samples, which are mainly composed by light atoms and preferentially live in a water environment, there is a particular interest in performing measurement in the so-called water window, i.e., the energy range between carbon (282 eV) and oxygen (533 eV) K-edge. In this range, the absorption contrast between the carbon of organelles and the water of both cytoplasm and the liquid surrounding the cell is quite high. For this reason, the EuPRAXIA@SPARC_LAB will be particularly suitable to perform high-contrast imaging experiments on biological samples in their living, hydrated, native state, without the need of cooling or staining them, as is the case for other microscopy techniques such as electron microscopy.

2.2. Time-Resolved X-ray Absorption Spectroscopy in the Water Window

Besides imaging experiments, which typically require single wavelength pulses, spectroscopy experiments requiring the scan of a (limited) energy range can also be performed at a FEL.

The main advantage of performing XAS experiments at the EuPRAXIA@SPARC_LAB FEL with respect to the more compact HGG sources (as, for example, those described in [25]) is the number of photons per pulse, which is foreseen to be as high as 10^{12} photons/pulse, thus being still significantly higher than that currently achievable at HGG sources (e.g., Popmintchev et al. [26] reported a fux of more than 10^9 photons/second in the water window energy range). The high intensity of the FEL pulses will allow acquiring data with a good signal-to-noise ratio from single-shots measurements.

In this context, X-ray Absorption Spectroscopy (XAS) can be used as a tool to directly observe the molecular structure during chemical dynamics studies [27,28]. Real time observations require indeed fast time and small spatial resolutions, which can be guaranteed by the short, intense EuPRAXIA@SPARC_LAB FEL pulses. In particular, either by tuning the undulators to the appropriate energy, or exploiting the natural jitter of the FEL radiation generated in SASE mode, the experiments performed at EuPRAXIA@SPARC_LAB will allow measuring the informative, low-energy portion of the XAS spectrum, the so-called XANES (X-ray Absorption Near Edge Spectroscopy) region. Quantitative analysis tools of XANES data are nowadays available [29,30] including those based on first principles calculations [31–33]. Therefore, FEL-XAS measurements will become a powerful tool to provide unique information on the local geometry, electron density and spin states around selected atomic moieties [34]. Soft X-rays as the ones that will be produced by EuPRAXIA@SPARC_LAB are well suited for chemical and biological studies in the water window region. This region includes the K edge of elements such as C, N and O, and the L edge of 3d transition metals, which are of interest in many biologically relevant cases [35,36].

Examples of pioneering soft X-ray L-edge FEL-XAS transmission experiments include measurements of Al, Ge and Ti thin films for variable fluence (see, for example, [37–39]). In those experiments, ultrafast electron heating pumping matter at extremely high temperatures, as well as saturable absorption effects were observed. FEL experiments were found to be extremely useful to explore highly uniform warm dense matter (WDM) conditions, a regime exceedingly difficult to reach in present laboratory studies, but relevant to various fields, including high-pressure and planetary science, astrophysics, and plasma production. Various FEL-based ultrafast techniques can be used to probe WDM properties at electron temperatures in the 1–10 eV range and beyond. Those previous results naturally call for further challenging experiments at the EuPRAXIA@SPARC_LAB FEL as well as for parallel developments of suitable interpretation schemes for modeling and understanding the X-ray absorption cross section under high-fluence conditions (see [40] and refs. therein).

For experiments near the chemically relevant carbon K edge at 284.2 eV, the EuPRAXIA@SPARC_LAB FEL can be used to study dissociation reactions of molecular cations, that until today could not been resolved in time, using transient absorption at the carbon K-edge. Moreover, XAS measurements at the

L-edge of 3d transition metals provides unique information on the local metal charge and spin states by directly probing 3d-derived molecular orbitals through 2p-3d transitions. However, this soft X-ray technique has been rarely used at synchrotron facilities for mechanistic studies of metalloenzymes due to the problems with X-ray-induced sample damage and strong background signals from light elements that can dominate the low metal signal. It has been recently shown [34] that Mn L-edge absorption spectra can be collected at room temperature at a FEL. This paves the way for future structural and dynamical studies of metalloenzymes exploiting soft X-ray FEL radiation such as the one produced by EuPRAXIA@SPARC_LAB.

2.3. Time-Resolved Coherent Raman Experiments with X-ray Pulses

One of the most intriguing challenges in modern scientific research is the capability of monitoring transient atomic motions that govern physical, chemical and biological phenomena, measuring structural molecular changes of reacting species over few Ångstrom lengths on sub-picosecond timescales. The standard approach used to investigate structural dynamics is the pump–probe scheme, in which light pulses are used to first excite (pump) and subsequently interrogate (probe) a system [41]. The use of intense, ultra-short, soft X-ray radiation pulses such as those generated by the EuPRAXIA@SPARC_LAB FEL would tremendously benefit pump–probe investigations, whereof two different situations will be addressed: on the one hand, X-ray pulses can be exploited as pump pulse for stimulating chemical reactions or for generating coherent excitations, and, on the other hand, they can be used as selective probe to monitor the evolution from reactant to photoproduct.

Raman spectroscopy is a very powerful experimental tool for the detection of molecular vibrations, which are related to the force constant between atoms. In this scenario, accessing the Raman spectrum during and upon the FEL interaction would disclose any vibrational and structural modification occurring on the system under investigation. The subsequent electronic relaxation modifies the force field, generating a fragmentation of the molecule. To follow the evolution from the point of view of molecular vibrations, two crucial requirements are needed: (1) collimated signal, to avoid the luminescence background generated from the sample after the FEL interaction; and (2) sub-picosecond time resolution to follow the fragmentation process. Therefore, spontaneous Raman spectroscopies are ineffective in this exploration, due to the isotropic signal and temporal resolution [42], compromised by the fundamental restrictions dictated by the Fourier transform limit. Femtosecond stimulated Raman scattering (FSRS) is a recently developed technique [43–47], in which a femtosecond actinic pulse (AP) initiates the photochemistry of interest. The system is subsequently interrogated by a pair of overlapped pulses: the joint presence of a broadband ultrashort probe pulse (PP) and a narrowband picosecond Raman pulse (RP) induces vibrational coherences which are read out as heterodyne coherent Raman signals [48,49]. Notably, the probed Raman features are engraved onto the highly directional PP, and, hence, SRS provides an efficient suppression of the incoherent fluorescence background. Moreover, thanks to the different temporal and spectral properties of the pulses, femtosecond SRS represents an ideal tool to study structural changes in ultrafast photophysical and photochemical processes, providing both femtosecond time precision and high spectral resolution [45,50–52]. The narrowband RP can be generated from a two-stage optical parametric amplifier that produces tunable infrared-visible pulses, followed by spectral compression via frequency doubling in a 25 mm beta-barium borate crystal [53]. The femtosecond AP, so far in the visible spectral region, can be replaced with XUV-FEL. In this way, tuning the soft X-ray wavelength in resonance with a specific atomic absorption edge, it would be possible to selectively excite specific atoms and follow the temporal evolution of Raman mode disappearance, which depends on atomic role in molecular oscillation and the electronic coupling between atoms. From another perspective, pulses in the X-ray domain, resonant with valence excited-state transitions, can be used as probe pulses in Raman based spectroscopies, enabling to selectively isolate contributions from specific sites of molecular moieties. In particular, combining an X-ray femtosecond probe pulse, with a visible photochemical pump pulse, would give the chance to perform X-ray Impulsive Vibrational Scattering (X-IVS), in which a visible pump pulse, besides triggering a photo-reaction, stimulates vibrational coherences on the system, modulating the transmission of a temporally delayed XUV probe, at the frequencies of the coherently activated vibrations [54,55]. For this reason, recording the transmission of the probe pulse enables real-time monitoring of Raman active modes. Fourier transforming the detected signal over the temporal delay between pump and probe recovers the transient Raman spectrum of the system under investigation. While IVS has been extensively exploited in the visible spectral region for probing ground and excited state coherences on both molecular and solid-state compounds [56,57], its potential in the X-ray domain is still an unexplored territory, which can be disclosed thanks to the EuPRAXIA@SPARC_LAB FEL experimental endstation. Notably, in close analogy with the atomic selectivity achieved by using a XUV pump pulse, employing a probe pulse resonant with a specific electronic transition absorption edge would provide the chance of isolating coherent atomic motions involving only the desired atomic moieties. A schematic view of a pump–probe Raman experiment is depicted in Figure 4.

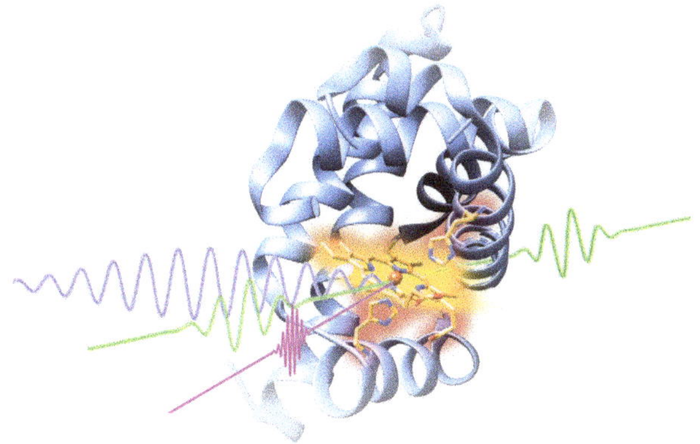

Figure 4. Schematic view of a time-resolved (pump–probe) Raman experiment on a protein.

Further development will be done in the field of localized dynamic studies by nano-Raman instruments. Both apertureless and fiber-based Scanning Near-field Optical Microscopy (SNOM) will be used to increase the lateral spatial resolution in the tens of nm: in this case, the excitation of the sample is kept as uniform as possible, and collection of scattered signal from local spot on the sample is conducted [58,59].

Another field where one can exploit the XUV photons generated by the EuPRAXIA@SPARC_LAB FEL is the study by means of coherent electronic Raman process [39] of photo-induced chemical processes represented by the detection of electronic coherences (based on a composite X-ray pulse sequence) generated during the system dynamics [60]. For example, within such a scheme, a combination of short, soft X-ray FEL pulses can be used to directly detect the passage through conical intersections (CIs). Notably, the photoinduced excited state dynamics of polyatomic molecules is often dominated by CI, regions of degeneracy between two or more electronic surfaces [61]. The dynamical behavior of molecules in the vicinity of CIs dictates the resulting photophysics and photochemistry of the molecule. Given the ubiquity and importance of CIs in all photoinduced processes, from solar energy conversion to vision, finding a direct, experimental observable of the dynamics through a CI would be a significant development in our understanding of excited state photo-induced dynamics.

2.4. Photo-Fragmentation of Molecules

Another kind of applications that would largely benefit from the peculiar features of the EuPRAXIA@SPARC_LAB FEL radiation is represented by the wide class of experiments aimed at studying the interaction of intense radiation pulses with molecules. How organic and biological molecules redistribute the energy of absorbed light is indeed a key fundamental question in organic chemistry and biology which time-resolved experiment can help to settle [62–67]. This class of experiments will help understanding the basic mechanisms of photo-protection/damage of amino acids [68], proteins and DNA/RNA [69]. XUV or X-rays pump laser pulses of low intensity with a few femtoseconds duration contain photons ranging up to hundreds of eV. Single-photon ionization is a dominant absorption channel triggering the ultrafast charge migration process in the parent cation. Probing the resulting non-equilibrium dynamics using short, intense pulses such as those produced by the EuPRAXIA@SPARC_LAB FEL at delays varying on the femtosecond timescale allows resolving in real-time the electron density through time-resolved imaging [70–74].

An example of what can be seen in a photo-fragmentation experiment as it will be implemented at EuPRAXIA@SPARC_LAB FEL is given in Figure 5. In particular, we display the real-space distribution of the molecular charge at six representative times of the phenylalanine amino acid after illumination with an ionizing XUV 300-as pulse.

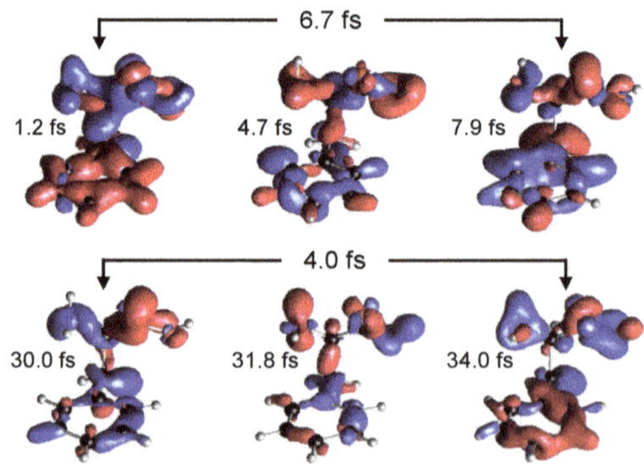

Figure 5. Real-space distribution of the molecular charge at six representative times of the phenylalanine amino acid after illumination with an ionizing XUV 300-as pulse. The figure is taken from Reference [68] (Copyright 2018 by the American Chemical Society).

2.5. Resonant Inelastic X-ray Scattering

In atomic inner shell spectroscopy, spectra can show peculiar characteristics associated with a variety of different scattering interactions. When atoms are irradiated with incident energy lower but close to an absorption edge, scattering peaks appear due to an inelastic process known as Resonant Inelastic X-ray Scattering (RIXS) or X-ray Resonant Raman Scattering [75]. These RIXS peaks display typical features, such as a characteristic long-tail spreading to the region of lower energy. This scattering process is a high demanding photon flux. In general, RIXS experiments are carried out at synchrotron facilities using high-resolution spectrometers for detecting the scattering signal. Nevertheless, in recent years, RIXS has been observed using an Energy Dispersive Setup (EDS) with synchrotron radiation. The analysis of the collected signal shows that hidden on the peak tails there is valuable information about the chemical environment of the atom under study [76]. During the last decade, several works have been published showing the first applications of a novel RIXS tool (named EDIXS) for the discrimination, determination and characterization of chemical environments in a variety of samples and irradiation geometries and even combined with other spectroscopic techniques [76–86]. Due to its versatility, EDIXS was applied in the typical 45°–45° setup for inspection of the material bulk, in total reflection for the study of the most external atomic layers of a sample, in grazing incidence used for depth resolved chemical speciation analysis and even in confocal setup to obtain chemical state information in a 3D regime, reaching nanometric spatial resolution. Owing to the EDIXS high sensitivity, this technique can be extended to the study of local chemical environments with applications in many field of science, as geology, chemistry, physics, material science and industry, etc., where a precise quantification of different compounds is required [79]. This methodology is fast, reliable and straightforward. It has several benefits compared with other spectroscopic techniques, such as fast acquisition, low self-absorption and the avoidance of any energy scan during the measurements. A remarkable field of application of EDIXS is in the context of pulsed X-ray sources, e.g., FELs. Besides the application of EDIXS for fast structural characterization of materials, this tool allows time-resolved investigations of a variety of atomic processes and of the dynamics of samples of interest exposed, for example, to changing conditions of temperature, atmosphere, pressure, etc. Previous results regarding time-resolved discrimination of chemical environments [80] have showed time resolution of the order of the second when monochromatic synchrotron radiation was used (flux $\sim 10^8$ ph/s). Due to the higher FEL photon flux, we expect to obtaining sub-second, and even millisecond, time-resolved experiments when a FEL source is used. This kind of (very) fast characterizations are currently impossible to achieve by conventional methods. Even non-conventional sources (storage rings) employing traditional techniques for atomic environment analysis (EXAFS, XANES, etc.) are useless in time-resolved spectroscopy because of the need of energy scan. This limitation establishes an ultimate frontier for these techniques that cannot be overcome during time dependent measurements. At this point, the one-shot character of EDIXS makes a crucial difference in favor of it. There are a variety of relevant cases to study with time-resolved EDIXS, both in basic research and in applications to the industry and technology fields. As for the multiple applications of time-resolved EDIXS using the EuPRAXIA@SPARC_LAB FEL (with produces photons with a maximum energy of $\sim$415 eV) as basic research, we mention the analysis of nitrogen and carbon states in the evolution of biological systems, for example in the study of the role of nitrogen during photosynthesis and of the chemical state of carbon during cell divisions. Concerning technological and industrial applications, a wide range of opportunities exists, since carbon plays a role in many situations. Just to mention a few of them: diamond structural variations under high conditions of pressure or graphene and complex carbon structures reactions to external excitations. The key element at the basis of the feasibility of all the RIXS experiments we have illustrated is the combination of a fast time resolution technique with the EuPRAXIA@SPARC_LAB source, delivering extremely high photon fluxes.

2.6. THz/MIR Sources

The interest in THz radiation is recognized since many years for its potential to advance research in several scientific fields. In addition, THz research has many industrial prospects, so that THz activities may offer potential spin-off not only associated to condensed matter basic research, e.g., semiconductor and superconductors materials, whose characterization may have a direct impact on many technologies, but also in R&D of detectors and imaging. A great expectation for industry is the development of imaging for biomedical applications and security issues.

THz radiation lies between the photonic and the electronic bands of the electromagnetic spectrum, and it extends from 300 GHz up to 10 THz. THz is non-ionizing and highly penetrating in a large variety of dielectric materials, e.g., plastic, ceramics, and paper. The wavelength of the THz radiation is of the order of many important physical, chemical and biological processes (see Figure 6), including superconducting gaps, exotic electronic transitions and protein dynamical processes. The THz part of the spectrum is energetically equivalent to many important physical, chemical and biological processes including superconducting gaps, exotic electronic transitions and protein dynamical processes. Recently, a new generation of sources, based on particle accelerators, allows increasing the average and peak power, by many orders of magnitude, and extends the spectral range up to the Mid-Infrared (100 THz, Middle-InfraRed (MIR)), making the whole spectral region accessible to different frequency- and time-domain experiments. Indeed, a linac-driven THz/MIR source can deliver broadband pulses with femtosecond shaping, and with the possibility to store a high energy in a single pulse [87]. In addition, taking advantage of electron beam manipulation techniques, high power, narrow-band radiation can be also generated [88]. Finally, high brightness electron beams also permit the possibility to extend the emission towards the MIR, having a unique source covering three decades in wavelength from 1000 microns to 1 micron. This provides a unique chance to realize THz/MIR-pump/THz/MIR probe spectroscopy, a technique essentially unexplored up to now.

The potential of THz and MIR frequency and time domains spectroscopies are displayed in Figure 6, where we show a not exhaustive review of excitations whose characteristic energy are in resonance with those of specific processes.

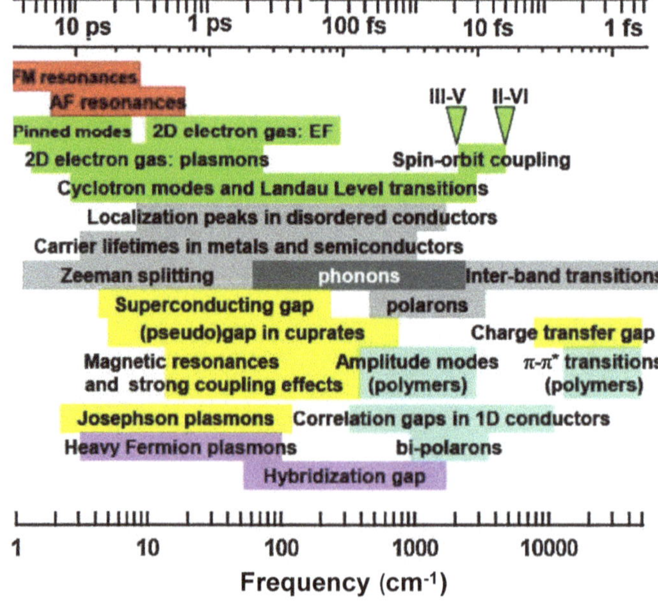

Figure 6. Frequency and time domain of THz/MIR spectroscopy.

An electromagnetic source can be characterized in terms of time duration, field strength, pulse shape, bandwidth and frequency. Their choice depends on the class of experiments of interest. An effective

THz/MIR source should have higher peak fields, from 100 kV/cm to 50 MV/cm, the coverage of a spectral range up to a frequency of 100 THz, a full pulse shaping and a sub-ps duration. The THz/MIR source at EuPRAXIA@SPARC_LAB will be designed to achieve these requirements.

THz/MIR radiation will be generated at EuPRAXIA@SPARC_LAB through different production schemes based on ultra-short, i.e., $\approx$10–100 fs, electron bunches. Two beam lines are considered in the first phase of the EuPRAXIA@SPARC_LAB project: one at low energy, i.e., 30–50 MeV, and the second one at higher energy, i.e., 1 GeV, in proximity of the FEL extraction site (see Figure 2).

Being produced by the same electron beam, these two sources are naturally synchronized on few femtosecond time scales. To perform THz pump X-ray probe experiments, we plan to take advantage of laser-based THz streaking, which effectively phase-locks a single-cycle THz pulse to the X-ray pulse. This technique simultaneously clocks the arrival time of the two sources and allows the measurement of the X-ray pulse temporal profile with a precision of tens of femtoseconds.

The first beamline, consisting of a THz/MIR SASE FEL able to emit quasi-monochromatic and fully polarized (with variable polarization), radiation from THz to MIR, will be optimized for experiments involving high peak power and narrow band THz/MIR radiation. The second beamline will combine the Coherent Diffraction Radiation [89], emitted from a rectangular slit in a metallic screen, to the VUV/X SASE FEL radiation to perform THz pump X-ray probe experiments.

Coherent Transition and Diffraction Radiation (CTR and CDR, respectively) are the chosen production mechanisms at high electron beam energy, i.e., $\sim$GeV scale, with the advantage of a broadband spectrum up to several THz depending on how short the bunch duration is. In the case of CDR, a further advantage is represented by the non-disruptiveness for the electron beam [90]. Both electron and THz representative radiation parameters are reported in Table 1.

Table 1. Electron beam and THz source parameters from CDR.

Beam Parameters		Source Parameters	
E (GeV)	1	Frequency (THz)	0.3–10
Q (pC)	200	P_{peak} (MW)	>100
σ_z (μm)	30	E_{ph} (μJ)	$\approx$100

At low electron beam energy, around 30 and 50 MeV, a SASE FEL operating in the MIR/THz range has been considered for the generation of highly intense narrow band, tunable radiation. A SPARC-like undulator [91] (2.8 cm period, K parameter ($K = \frac{eB_0\lambda_u}{2\pi mc}$, with B_0 the magnetic field on axis and λ_u the undulator period; e and m are the electron charge and mass, and c the speed of light in vacuum) of 1.2) with variable gap has been considered for the calculation of MIR/THz radiation based on GPT simulations [92] for the electron beam dynamics and Ming Xie formulas [93] for the SASE FEL performances; saturation occurs within 5 m of undulator length. Both electron beam and MIR/THz radiation parameters are listed in Table 2.

Table 2. Electron beam and MIR/THz source parameters from a MIR SASE FEL.

Beam Parameters		Source Parameters	
E (MeV)	30–50	λ_r (μm)	10–3
Q (pC)	200	L_{sat}	3–4.4
σ_z (μm)	50	P_{sat} (MW)	140–135
I_{peak} (kA)	480	N_{ph}	$\sim$10^{15}
$\Delta E/E$ (%)	0.1–0.4	E_{ph} (μJ)	$\sim$60

The magnetic structure of the undulator will be optimized to provide fully polarized light. The polarization can be modfied from linear, circular, elliptical, to more complex structures such as helicoidal polarization. These polarization states, which are absolutely unconventional for THz and MIR lights and of difficult realization with thermal- and laser-based sources, will provide the possibility to pump exotic modes such as skyrmions in magnetic systems, Weyl and Dirac fermions [94] in non-trivial quantum matter, and Higgs and Legett modes in multi-gap superconductors and charge-density wave materials. Experiments in which all phonon modes of exotic systems can be selectively pumped will be also accessible, opening the possibility to control the lattice structure on the ps-scale and, consequently, to modulate the electronic ground state of the systems [95]. Localized dinamical spectroscopic imaging will be performed by THz apertureless-SNOM, where the tip's antenna function will allow performing imaging with a lateral resolution well below 1 micron [96].

3. Discussion and Conclusions

In this paper, we summarize the main experimental lines of investigation that can be implemented at the EuPRAXIA@SPARC_LAB FEL exploiting its ultra-short, bright FEL pulses generated in the "water-window". The realization of the EuPRAXIA@SPARC_LAB infrastructure will allow INFN to consolidate a strong scientific, technological and industrial role in a competing international context. To exploit at best the features of this compact machine, a great effort will have to be addressed in developing, designing and assembling all the optical components necessary to deliver the FEL photons to the user endstation and in developing and characterizing detectors able to optimize the signal-to-noise ratio for all the foreseen classes of experiments. We are confident that the EuPRAXIA@SPARC_LAB photon source, with its multi-purpose beamline, designed and equipped to perform the classes of experiments highlighted in this paper, will be highly beneficial for the national and international community of FEL radiation users.

Author Contributions: Conceptualization, writing—review and editing, all authors; and Writing—original draft preparation, all authors.

Funding: This research was funded by European Union's Horizon 2020 research and innovation programme under grant agreement No. 653782. One of the authors (S.D.) acknowledges the support by the Competitiveness Program of NRNU MEPhI (Moscow).

Conflicts of Interest: The authors declare no conflict of interest.

Abbreviations

The following abbreviations are used in this manuscript:

AP	Actinic Pulse
CDI	Coherent Diffraction Imaging
CDR	Coherent Diffraction Radiation
CTR	Coherent Transition Radiation
CI	Conical Intersection
EDS	Energy Dispersive Setup
FEL	Free Electron Laser
FSRS	Femtosecond stimulated Raman scattering
MIR	Middle-InfraRed
PP	Probe Pulse
RIXS	Resonant Inelastic X-ray Scattering
RP	Raman Pulse

SASE   Self Amplification of SpontanEous radiation
SNOM   Scanning Near-field Optical Microscopy
THz   Tera Hertz
XAS   X-ray Absortpion Spectroscopy
X-RIVS   X-ray Impulsive Vibrational Scattering
XUV   eXtreme Ultra Violet

References

1. Chapman, H.N.; Fromme, P.; Barty, A.; White, T.A.; Kirian, R.A.; Aquila, A.; Hunter, M.S.; Schulz, J.; DePonte, D.P.; Weierstall, U.; et al. Femtosecond X-ray protein nanocrystallography. *Nature* **2011**, *470*, 73–77. [CrossRef] [PubMed]

2. Boutet, S.; Lomb, L.; Williams, G.J.; Barends, T.R.; Aquila, A.; Doak, R.B.; Weierstall, U.; DePonte, D.P.; Steinbrener, J.; Shoeman, R.L.; et al. High-resolution protein structure determination by serial femtosecond crystallography. *Science* **2012**, *337*, 362–364. [CrossRef] [PubMed]

3. Seibert, M.M.; Ekeberg, T.; Maia, F.R.; Svenda, M.; Andreasson, J.; Jönsson, O.; Odić, D.; Iwan, B.; Rocker, A.; Westphal, D.; et al. Single mimivirus particles intercepted and imaged with an X-ray laser. *Nature* **2011**, *470*, 78–81. [CrossRef] [PubMed]

4. Hantke, M.F.; Hasse, D.; Maia, F.R.; Ekeberg, T.; John, K.; Svenda, M.; Loh, N.D.; Martin, A.V.; Timneanu, N.; Larsson, D.S.; et al. High-throughput imaging of heterogeneous cell organelles with an X-ray laser. *Nat. Photonics* **2014**, *8*, 943–949. [CrossRef]

5. Van Der Schot, G.; Svenda, M.; Maia, F.R.; Hantke, M.; DePonte, D.P.; Seibert, M.M.; Aquila, A.; Schulz, J.; Kirian, R.; Liang, M.; et al. Imaging single cells in a beam of live cyanobacteria with an X-ray laser. *Nat. Commun.* **2015**, *6*, 5704. [CrossRef] [PubMed]

6. Fan, J.; Sun, Z.; Wang, Y.; Park, J.; Kim, S.; Gallagher-Jones, M.; Kim, Y.; Song, C.; Yao, S.; Zhang, J.; et al. Single-pulse enhanced coherent diffraction imaging of bacteria with an X-ray free-electron laser. *Sci. Rep.* **2016**, *6*, 34008. [CrossRef] [PubMed]

7. Henderson, R. The potential and limitations of neutrons, electrons and X-rays for atomic resolution microscopy of unstained biological molecules. *Q. Rev. Biophys.* **1995**, *28*, 171–193. [CrossRef] [PubMed]

8. Gutt, C.; Streit-Nierobisch, S.; Stadler, L.M.; Pfau, B.; Günther, C.; Könnecke, R.; Frömter, R.; Kobs, A.; Stickler, D.; Oepen, H.; et al. Single-pulse resonant magnetic scattering using a soft X-ray free-electron laser. *Phys. Rev. B* **2010**, *81*, 100401. [CrossRef]

9. Ferrario, M.; Alesini, D.; Anania, M.; Artioli, M.; Bacci, A.; Bartocci, S.; Bedogni, R.; Bellaveglia, M.; Biagioni, A.; Bisesto, F.; et al. EuPRAXIA@ SPARC_LAB Design study towards a compact FEL facility at LNF. *Nuclear Instrum. Methods Phys. Res. Sect. A* **2018**, *909*, 134–138. [CrossRef]

10. Villa, F.; Cianchi, A.; Coreno, M.; Dabagov, S.; Marcelli, A.; Minicozzi, V.; Morante, S.; Stellato, F. Design study of a photon beamline for a soft X-ray FEL driven by high gradient acceleration at EuPRAXIA@ SPARC_LAB. *Nuclear Instrum. Methods Phys. Res. Sect. A* **2018**, *909*, 294–297. [CrossRef]

11. Oberthür, D. Biological single-particle imaging using XFELs–towards the next resolution revolution. *IUCrJ* **2018**, *5*, 663. [CrossRef] [PubMed]

12. Reddy, H.K.; Yoon, C.H.; Aquila, A.; Awel, S.; Ayyer, K.; Barty, A.; Berntsen, P.; Bielecki, J.; Bobkov, S.; Bucher, M.; et al. Coherent soft X-ray diffraction imaging of coliphage PR772 at the Linac coherent light source. *Sci. Data* **2017**, *4*, 170079. [CrossRef] [PubMed]

13. Rose, M.; Bobkov, S.; Ayyer, K.; Kurta, R.P.; Dzhigaev, D.; Kim, Y.Y.; Morgan, A.J.; Yoon, C.H.; Westphal, D.; Bielecki, J.; et al. Single-particle imaging without symmetry constraints at an X-ray free-electron laser. *IUCrJ* **2018**, *5*, 727–736. [CrossRef] [PubMed]

14. Munke, A.; Andreasson, J.; Aquila, A.; Awel, S.; Ayyer, K.; Barty, A.; Bean, R.J.; Berntsen, P.; Bielecki, J.; Boutet, S.; et al. Coherent diffraction of single Rice Dwarf virus particles using hard X-rays at the Linac Coherent Light Source. *Sci. Data* **2016**, *3*, 160064. [CrossRef] [PubMed]

15. Takayama, Y.; Inui, Y.; Sekiguchi, Y.; Kobayashi, A.; Oroguchi, T.; Yamamoto, M.; Matsunaga, S.; Nakasako, M. Coherent X-ray diffraction imaging of chloroplasts from Cyanidioschyzon merolae by using X-ray free electron laser. *Plant Cell Physiol.* **2015**, *56*, 1272–1286. [CrossRef] [PubMed]

16. Van Der Schot, G.; Svenda, M.; Maia, F.R.; Hantke, M.F.; DePonte, D.P.; Seibert, M.M.; Aquila, A.; Schulz, J.; Kirian, R.A.; Liang, M.; et al. Open data set of live cyanobacterial cells imaged using an X-ray laser. *Sci. Data* **2016**, *3*, 160058. [CrossRef] [PubMed]

17. Hantke, M.F.; Hasse, D.; Ekeberg, T.; John, K.; Svenda, M.; Loh, D.; Martin, A.V.; Timneanu, N.; Larsson, D.S.; Van Der Schot, G.; et al. A data set from flash X-ray imaging of carboxysomes. *Sci. Data* **2016**, *3*, 160061. [CrossRef] [PubMed]

18. Ackermann, W.a.; Asova, G.; Ayvazyan, V.; Azima, A.; Baboi, N.; Bähr, J.; Balandin, V.; Beutner, B.; Brandt, A.; Bolzmann, A.; et al. Operation of a free-electron laser from the extreme ultraviolet to the water window. *Nat. Photonics* **2007**, *1*, 336. [CrossRef]

19. Popp, D.; Loh, N.D.; Zorgati, H.; Ghoshdastider, U.; Liow, L.T.; Ivanova, M.I.; Larsson, M.; DePonte, D.P.; Bean, R.; Beyerlein, K.R.; et al. Flow-aligned, single-shot fiber diffraction using a femtosecond X-ray free-electron laser. *Cytoskeleton* **2017**, *74*, 472–481. [CrossRef] [PubMed]

20. Stellato, F.; Fusco, Z.; Chiaraluce, R.; Consalvi, V.; Dinarelli, S.; Placidi, E.; Petrosino, M.; Rossi, G.; Minicozzi, V.; Morante, S. The effect of β-sheet breaker peptides on metal associated Amyloid-β peptide aggregation process. *Biophys. Chem.* **2017**, *229*, 110–114. [CrossRef] [PubMed]

21. Carbonaro, M.; Di Venere, A.; Filabozzi, A.; Maselli, P.; Minicozzi, V.; Morante, S.; Nicolai, E.; Nucara, A.; Placidi, E.; Stellato, F. Role of dietary antioxidant (-)-epicatechin in the development of β-lactoglobulin fibrils. *Biochim. Biophys. Acta* **2016**, *1864*, 766–772. [CrossRef] [PubMed]

22. Carbonaro, M.; Ripanti, F.; Filabozzi, A.; Minicozzi, V.; Stellato, F.; Placidi, E.; Morante, S.; Di Venere, A.; Nicolai, E.; Postorino, P.; et al. Human insulin fibrillogenesis in the presence of epigallocatechin gallate and melatonin: Structural insights from a biophysical approach. *Int. J. Biol. Macromol.* **2018**, *115*, 1157–1164. [CrossRef] [PubMed]

23. Ekeberg, T.; Svenda, M.; Abergel, C.; Maia, F.R.; Seltzer, V.; Claverie, J.M.; Hantke, M.; Jönsson, O.; Nettelblad, C.; Van Der Schot, G.; et al. Three-dimensional reconstruction of the giant mimivirus particle with an X-ray free-electron laser. *Phys. Rev. Lett.* **2015**, *114*, 098102. [CrossRef] [PubMed]

24. Hantke, M.F.; Ekeberg, T.; Maia, F.R. Condor: A simulation tool for flash X-ray imaging. *J. Appl. Crystallogr.* **2016**, *49*, 1356–1362. [CrossRef] [PubMed]

25. Pertot, Y.; Schmidt, C.; Matthews, M.; Chauvet, A.; Huppert, M.; Svoboda, V.; von Conta, A.; Tehlar, A.; Baykusheva, D.; Wolf, J.P.; et al. Time-resolved X-ray absorption spectroscopy with a water window high-harmonic source. *Science* **2017**, *355*, 264–267. [CrossRef] [PubMed]

26. Popmintchev, D.; Galloway, B.R.; Chen, M.C.; Dollar, F.; Mancuso, C.A.; Hankla, A.; Miaja-Avila, L.; O'Neil, G.; Shaw, J.M.; Fan, G.; et al. Near-and extended-edge X-ray-absorption fine-structure spectroscopy using ultrafast coherent high-order harmonic supercontinua. *Phys. Rev. Lett.* **2018**, *120*, 093002. [CrossRef] [PubMed]

27. Chergui, M.; Zewail, A.H. Electron and X-ray Methods of Ultrafast Structural Dynamics: Advances and Applications. *ChemPhysChem* **2009**, *10*, 28–43. [CrossRef] [PubMed]

28. Bressler, C.; Chergui, M. Molecular structural dynamics probed by ultrafast X-ray absorption spectroscopy. *Annu. Rev. Phys. Chem.* **2010**, *61*, 263–282. [CrossRef] [PubMed]

29. Benfatto, M.; Della Longa, S.; Natoli, C. The MXAN procedure: A new method for analysing the XANES spectra of metalloproteins to obtain structural quantitative information. *J. Synchrotron Radiat.* **2003**, *10*, 51–57. [CrossRef] [PubMed]

30. Bunau, O.; Joly, Y. Self-consistent aspects of X-ray absorption calculations. *J. Phys. Condens. Matter* **2009**, *21*, 345501. [CrossRef] [PubMed]

31. Rehr, J.J.; Kas, J.J.; Vila, F.D.; Prange, M.P.; Jorissen, K. Parameter-free calculations of X-ray spectra with FEFF9. *Phys. Chem. Chem. Phys.* **2010**, *12*, 5503–5513. [CrossRef] [PubMed]

32. La Penna, G.; Minicozzi, V.; Morante, S.; Rossi, G.; Stellato, F. A first-principle calculation of the XANES spectrum of Cu^{2+} in water. *J. Chem. Phys.* **2015**, *143*, 124508. [CrossRef] [PubMed]

33. Stellato, F.; Calandra, M.; D'Acapito, F.; De Santis, E.; La Penna, G.; Rossi, G.; Morante, S. Multi-scale theoretical approach to X-ray absorption spectra in disordered systems: An application to the study of Zn (II) in water. *Phys. Chem. Chem. Phys.* **2018**, *20*, 24775–24782. [CrossRef] [PubMed]
34. Kubin, M.; Kern, J.; Gul, S.; Kroll, T.; Chatterjee, R.; Löchel, H.; Fuller, F.D.; Sierra, R.G.; Quevedo, W.; Weniger, C.; et al. Soft X-ray absorption spectroscopy of metalloproteins and high-valent metal-complexes at room temperature using free-electron lasers. *Struct. Dyn.* **2017**, *4*, 054307. [CrossRef] [PubMed]
35. De Santis, E.; Minicozzi, V.; Proux, O.; Rossi, G.; Silva, K.I.; Lawless, M.J.; Stellato, F.; Saxena, S.; Morante, S. Cu (II)–Zn (II) Cross-Modulation in Amyloid–Beta Peptide Binding: An X-ray Absorption Spectroscopy Study. *J. Phys. Chem. B* **2015**, *119*, 15813–15820. [CrossRef] [PubMed]
36. De Santis, E.; Shardlow, E.; Stellato, F.; Proux, O.; Rossi, G.; Exley, C.; Morante, S. X-ray Absorption Spectroscopy Measurements of Cu-ProIAPP Complexes at Physiological Concentrations. *Condens. Matter* **2019**, *4*, 13. [CrossRef]
37. Nagler, B.; Zastrau, U.; Fäustlin, R.R.; Vinko, S.M.; Whitcher, T.; Nelson, A.; Sobierajski, R.; Krzywinski, J.; Chalupsky, J.; Abreu, E.; et al. Turning solid aluminium transparent by intense soft X-ray photoionization. *Nat. Phys.* **2009**, *5*, 693.
38. Di Cicco, A.; Hatada, K.; Giangrisostomi, E.; Gunnella, R.; Bencivenga, F.; Principi, E.; Masciovecchio, C.; Filipponi, A. Interplay of electron heating and saturable absorption in ultrafast extreme ultraviolet transmission of condensed matter. *Phys. Rev. B* **2014**, *90*, 220303. [CrossRef]
39. Principi, E.; Giangrisostomi, E.; Cucini, R.; Bencivenga, F.; Battistoni, A.; Gessini, A.; Mincigrucci, R.; Saito, M.; Di Fonzo, S.; D'Amico, F.; et al. Free electron laser-driven ultrafast rearrangement of the electronic structure in Ti. *Struct. Dyn.* **2016**, *3*, 023604. [CrossRef] [PubMed]
40. Hatada, K.; Di Cicco, A. Modeling Non-Equilibrium Dynamics and Saturable Absorption Induced by Free Electron Laser Radiation. *Appl. Sci.* **2017**, *7*, 814. [CrossRef]
41. Zewail, A.H. Femtochemistry. Past, present, and future. *Pure Appl. Chem.* **2000**, *72*, 2219–2231. [CrossRef]
42. Kruglik, S.G.; Jasaitis, A.; Hola, K.; Yamashita, T.; Liebl, U.; Martin, J.L.; Vos, M.H. Subpicosecond oxygen trapping in the heme pocket of the oxygen sensor FixL observed by time-resolved resonance Raman spectroscopy. *Proc. Natl. Acad. Sci. USA* **2007**, *104*, 7408–7413. [CrossRef] [PubMed]
43. Batignani, G.; Pontecorvo, E.; Ferrante, C.; Aschi, M.; Elles, C.G.; Scopigno, T. Visualizing Excited-State Dynamics of a Diaryl Thiophene: Femtosecond Stimulated Raman Scattering as a Probe of Conjugated Molecules. *J. Phys. Chem. Lett.* **2016**, *7*, 2981–2988. [CrossRef] [PubMed]
44. Batignani, G.; Bossini, D.; Di Palo, N.; Ferrante, C.; Pontecorvo, E.; Cerullo, G.; Kimel, A.; Scopigno, T. Probing ultrafast photo-induced dynamics of the exchange energy in a Heisenberg antiferromagnet. *Nat. Photonics* **2015**, *9*, 506. [CrossRef]
45. Dietze, D.R.; Mathies, R.A. Femtosecond stimulated Raman spectroscopy. *ChemPhysChem* **2016**, *17*, 1224–1251. [CrossRef] [PubMed]
46. Ferrante, C.; Pontecorvo, E.; Cerullo, G.; Vos, M.; Scopigno, T. Direct observation of subpicosecond vibrational dynamics in photoexcited myoglobin. *Nat. Chem.* **2016**, *8*, 1137. [CrossRef] [PubMed]
47. McCamant, D.W.; Kukura, P.; Yoon, S.; Mathies, R.A. Femtosecond broadband stimulated Raman spectroscopy: Apparatus and methods. *Rev. Sci. Instrum.* **2004**, *75*, 4971–4980. [CrossRef] [PubMed]
48. Yoshizawa, M.; Hattori, Y.; Kobayashi, T. Femtosecond time-resolved resonance Raman gain spectroscopy in polydiacetylene. *Phys. Rev. B* **1994**, *49*, 13259. [CrossRef]
49. Ferrante, C.; Batignani, G.; Fumero, G.; Pontecorvo, E.; Virga, A.; Montemiglio, L.; Cerullo, G.; Vos, M.; Scopigno, T. Resonant broadband stimulated Raman scattering in myoglobin. *J. Raman Spectrosc.* **2018**, *49*, 913–920. [CrossRef]
50. Fumero, G.; Batignani, G.; Dorfman, K.E.; Mukamel, S.; Scopigno, T. On the Resolution Limit of Femtosecond Stimulated Raman Spectroscopy: Modelling Fifth-Order Signals with Overlapping Pulses. *ChemPhysChem* **2015**, *16*, 3438–3443. [CrossRef] [PubMed]
51. Kukura, P.; McCamant, D.W.; Mathies, R.A. Femtosecond stimulated Raman spectroscopy. *Annu. Rev. Phys. Chem.* **2007**, *58*, 461–488. [CrossRef] [PubMed]

52. Batignani, G.; Fumero, G.; Pontecorvo, E.; Ferrante, C.; Mukamel, S.; Scopigno, T. Genuine dynamics vs. cross phase modulation artefacts in Femtosecond Stimulated Raman Spectroscopy. *ACS Photonics* **2019**, *6*, 492–500. [CrossRef]

53. Pontecorvo, E.; Kapetanaki, S.; Badioli, M.; Brida, D.; Marangoni, M.; Cerullo, G.; Scopigno, T. Femtosecond stimulated Raman spectrometer in the 320–520 nm range. *Opt. Express* **2011**, *19*, 1107–1112. [CrossRef] [PubMed]

54. Monacelli, L.; Batignani, G.; Fumero, G.; Ferrante, C.; Mukamel, S.; Scopigno, T. Manipulating impulsive stimulated raman spectroscopy with a chirped probe pulse. *J. Phys. Chem. Lett.* **2017**, *8*, 966–974. [CrossRef] [PubMed]

55. Liebel, M.; Schnedermann, C.; Wende, T.; Kukura, P. Principles and applications of broadband impulsive vibrational spectroscopy. *J. Phys. Chem. A* **2015**, *119*, 9506–9517. [CrossRef] [PubMed]

56. Kuramochi, H.; Takeuchi, S.; Yonezawa, K.; Kamikubo, H.; Kataoka, M.; Tahara, T. Probing the early stages of photoreception in photoactive yellow protein with ultrafast time-domain Raman spectroscopy. *Nat. Chem.* **2017**, *9*, 660. [CrossRef] [PubMed]

57. Batignani, G.; Fumero, G.; Kandada, A.R.S.; Cerullo, G.; Gandini, M.; Ferrante, C.; Petrozza, A.; Scopigno, T. Probing femtosecond lattice displacement upon photo-carrier generation in lead halide perovskite. *Nat. Commun.* **2018**, *9*, 1971. [CrossRef] [PubMed]

58. Zavalin, A.; Cricenti, A.; Generosi, R.; Luce, M.; Morgan, S.; Piston, D. Nano-Raman mapping of porous glass ceramics with a scanning near-field optical microscope in collection mode. *Appl. Phys. Lett.* **2006**, *88*, 133126. [CrossRef]

59. Sonntag, M.D.; Pozzi, E.A.; Jiang, N.; Hersam, M.C.; Van Duyne, R.P. Recent advances in tip-enhanced Raman spectroscopy. *J. Phys. Chem. Lett.* **2014**, *5*, 3125–3130. [CrossRef] [PubMed]

60. Kowalewski, M.; Bennett, K.; Dorfman, K.E.; Mukamel, S. Catching conical intersections in the act: Monitoring transient electronic coherences by attosecond stimulated X-ray Raman signals. *Phys. Rev. Lett.* **2015**, *115*, 193003. [CrossRef] [PubMed]

61. Polli, D.; Altoè, P.; Weingart, O.; Spillane, K.M.; Manzoni, C.; Brida, D.; Tomasello, G.; Orlandi, G.; Kukura, P.; Mathies, R.A.; et al. Conical intersection dynamics of the primary photoisomerization event in vision. *Nature* **2010**, *467*, 440. [CrossRef] [PubMed]

62. Calegari, F.; Ayuso, D.; Trabattoni, A.; Belshaw, L.; De Camillis, S.; Anumula, S.; Frassetto, F.; Poletto, L.; Palacios, A.; Decleva, P.; et al. Ultrafast electron dynamics in phenylalanine initiated by attosecond pulses. *Science* **2014**, *346*, 336–339. [CrossRef] [PubMed]

63. Lara-Astiaso, M.; Galli, M.; Trabattoni, A.; Palacios, A.; Ayuso, D.; Frassetto, F.; Poletto, L.; De Camillis, S.; Greenwood, J.; Decleva, P.; et al. Attosecond Pump–Probe Spectroscopy of Charge Dynamics in Tryptophan. *J. Phys. Chem. Lett.* **2018**, *9*, 4570–4577. [CrossRef] [PubMed]

64. Beaulieu, S.; Comby, A.; Clergerie, A.; Caillat, J.; Descamps, D.; Dudovich, N.; Fabre, B.; Géneaux, R.; Légaré, F.; Petit, S.; et al. Attosecond-resolved photoionization of chiral molecules. *Science* **2017**, *358*, 1288–1294. [CrossRef] [PubMed]

65. Attar, A.R.; Bhattacherjee, A.; Pemmaraju, C.; Schnorr, K.; Closser, K.D.; Prendergast, D.; Leone, S.R. Femtosecond X-ray spectroscopy of an electrocyclic ring-opening reaction. *Science* **2017**, *356*, 54–59. [CrossRef] [PubMed]

66. Marciniak, A.; Despré, V.; Barillot, T.; Rouzée, A.; Galbraith, M.; Klei, J.; Yang, C.H.; Smeenk, C.; Loriot, V.; Reddy, S.N.; et al. XUV excitation followed by ultrafast non-adiabatic relaxation in PAH molecules as a femto-astrochemistry experiment. *Nat. Commun.* **2015**, *6*, 7909. [CrossRef] [PubMed]

67. Nelson, T.R.; Ondarse-Alvarez, D.; Oldani, N.; Rodriguez-Hernandez, B.; Alfonso-Hernandez, L.; Galindo, J.F.; Kleiman, V.D.; Fernandez-Alberti, S.; Roitberg, A.E.; Tretiak, S. Coherent exciton-vibrational dynamics and energy transfer in conjugated organics. *Nat. Commun.* **2018**, *9*, 2316. [CrossRef] [PubMed]

68. Perfetto, E.; Sangalli, D.; Marini, A.; Stefanucci, G. Ultrafast Charge Migration in XUV Photoexcited Phenylalanine: A First-Principles Study Based on Real-Time Nonequilibrium Green's Functions. *J. Phys. Chem. Lett.* **2018**, *9*, 1353–1358. [CrossRef] [PubMed]

69. Ren, X.; Wang, E.; Skitnevskaya, A.D.; Trofimov, A.B.; Gokhberg, K.; Dorn, A. Experimental evidence for ultrafast intermolecular relaxation processes in hydrated biomolecules. *Nat. Phys.* **2018**, *14*, 1062. [CrossRef]

70. Pullen, M.G.; Wolter, B.; Le, A.T.; Baudisch, M.; Hemmer, M.; Senftleben, A.; Schröter, C.D.; Ullrich, J.; Moshammer, R.; Lin, C.D.; et al. Imaging an aligned polyatomic molecule with laser-induced electron diffraction. *Nat. Commun.* **2015**, *6*, 7262. [CrossRef] [PubMed]

71. Erk, B.; Boll, R.; Trippel, S.; Anielski, D.; Foucar, L.; Rudek, B.; Epp, S.W.; Coffee, R.; Carron, S.; Schorb, S.; et al. Imaging charge transfer in iodomethane upon X-ray photoabsorption. *Science* **2014**, *345*, 288–291. [CrossRef] [PubMed]

72. Yang, J.; Zhu, X.; Wolf, T.J.; Li, Z.; Nunes, J.P.F.; Coffee, R.; Cryan, J.P.; Gühr, M.; Hegazy, K.; Heinz, T.F.; et al. Imaging CF3I conical intersection and photodissociation dynamics with ultrafast electron diffraction. *Science* **2018**, *361*, 64–67. [CrossRef] [PubMed]

73. Kraus, P.M.; Zürch, M.; Cushing, S.K.; Neumark, D.M.; Leone, S.R. The ultrafast X-ray spectroscopic revolution in chemical dynamics. *Nature* **2018**, *2*, 82–94. [CrossRef]

74. Kuleff, A.I.; Kryzhevoi, N.V.; Pernpointner, M.; Cederbaum, L.S. Core ionization initiates subfemtosecond charge migration in the valence shell of molecules. *Phys. Rev. Lett.* **2016**, *117*, 093002. [CrossRef] [PubMed]

75. Karydas, A.; Paradellis, T. Measurement of KL and LM resonant Raman scattering cross sections with a proton-induced X-ray beam. *J. Phys. B* **1997**, *30*, 1893. [CrossRef]

76. Leani, J.J.; Sánchez, H.J.; Valentinuzzi, M.; Pérez, C. Determination of the oxidation state by resonant-Raman scattering spectroscopy. *J. Anal. At. Spectrom.* **2011**, *26*, 378–382. [CrossRef]

77. Robledo, J.I.; Leani, J.J.; Karydas, A.G.; Migliori, A.; Pérez, C.A.; Sánchez, H.J. Energy-Dispersive Total-Reflection Resonant Inelastic X-ray Scattering as a Tool for Elemental Speciation in Contaminated Water. *Anal. Chem.* **2018**, *90*, 3886–3891. [CrossRef] [PubMed]

78. Leani, J.J.; Pérez, R.D.; Robledo, J.I.; Sánchez, H. 3D-reconstruction of chemical state distributions in stratified samples by spatially resolved micro-X-ray resonant Raman spectroscopy. *J. Anal. At. Spectrom.* **2017**, *32*, 402–407. [CrossRef]

79. Leani, J.J.; Robledo, J.I.; Sánchez, H.J. Quantitative speciation of manganese oxide mixtures by RIXS/RRS spectroscopy. *X-ray Spectrom.* **2017**, *46*, 507–511. [CrossRef]

80. Robledo, J.I.; Saánchez, H.J.; Leani, J.J.; Pérez, C.A. Exploratory methodology for retrieving oxidation state information from X-ray resonant Raman scattering spectrometry. *Anal. Chem.* **2015**, *87*, 3639–3645. [CrossRef] [PubMed]

81. Leani, J.J.; Sánchez, H.J.; Pérez, C.A. Oxide nanolayers in stratified samples studied by X-ray resonant Raman scattering at grazing incidence. *J. Spectrosc.* **2015**, *2015*, 618279. [CrossRef]

82. Sanchez, H.J.; Leani, J.J.; Pérez, C.; Pèrez, R.D. Arsenic Speciation by X-ray Spectroscopy using Resonant Raman Scattering. *J. Appl. Spectrosc.* **2014**, *80*, 912–916. [CrossRef]

83. Leani, J.J.; Saánchez, H.J.; Pérez, R.D.; Peérez, C. Depth profiling nano-analysis of chemical environments using resonant Raman spectroscopy at grazing incidence conditions. *Anal. Chem.* **2013**, *85*, 7069–7075. [CrossRef] [PubMed]

84. Leani, J.J.; Sánchez, H.J.; Valentinuzzi, M.C.; Pérez, C.; Grenón, M. Qualitative microanalysis of calcium local structure in tooth layers by means of micro-RRS. *J. Microsc.* **2013**, *250*, 111–115. [CrossRef] [PubMed]

85. Leani, J.J.; Sánchez, H.; Valentinuzzi, M.; Pérez, C. Chemical environment determination of iron oxides using RRS spectroscopy. *X-ray Spectrom.* **2011**, *40*, 254–256. [CrossRef]

86. Leani, J.J.; Robledo, J.I.; Sánchez, H.J. Energy dispersive inelastic X-ray scattering (EDIXS) spectroscopy–A review. *Spectrochim. Acta Part B* **2019**, *154*, 10–24. [CrossRef]

87. Chiadroni, E.; Bacci, A.; Bellaveglia, M.; Boscolo, M.; Castellano, M.; Cultrera, L.; Di Pirro, G.; Ferrario, M.; Ficcadenti, L.; Filippetto, D.; et al. The SPARC linear accelerator based terahertz source. *Appl. Phys. Lett.* **2013**, *102*, 094101. [CrossRef]

88. Giorgianni, F.; Anania, M.P.; Bellaveglia, M.; Biagioni, A.; Chiadroni, E.; Cianchi, A.; Daniele, M.; Del Franco, M.; Di Giovenale, D.; Di Pirro, G.; et al. Tailoring of highly intense thz radiation through high brightness electron beams longitudinal manipulation. *Appl. Sci.* **2016**, *6*, 56. [CrossRef]

89.	Castellano, M.; Verzilov, V.; Catani, L.; Cianchi, A.; Orlandi, G.; Geitz, M. Measurements of coherent diffraction radiation and its application for bunch length diagnostics in particle accelerators. *Phys. Rev. E* **2001**, *63*, 056501. [CrossRef] [PubMed]
90.	Chiadroni, E. Bunch Length Characterization at the TTF VUV-FEL. Ph.D. Thesis, Università degli Studi di Roma "Tor Vergata", Roma, Italy, September 2006.
91.	Alesini, D.; Vicario, C.; Bertolucci, S.; Biagini, M.E.; Biscari, C.; Boni, R.; Boscolo, M.; Castellano, M.; Clozza, A.; Di Pirro, G.; et al. *Technical Design Report for the SPARC Advanced Photo-Injector*; Palumbo, L., Rosenzweig, J.B., Eds.; INFN: Pisa, Italy, 2004.
92.	Van der Geer, S.; De Loos, M.; Bongerd, D. General Particle Tracer: A 3D code for accelerator and beam line design. In Proceedings of the 5th European Particle Accelerator Conference, Stockholm, Sweden, 10–14 June 1996.
93.	Xie, M. Design optimization for an X-ray free electron laser driven by SLAC linac. In Proceedings of the IEEE Particle Accelerator Conference, Dallas, TX, USA, 1–5 May 1995; Volume 1, pp. 183–185.
94.	Giorgianni, F.; Chiadroni, E.; Rovere, A.; Cestelli-Guidi, M.; Perucchi, A.; Bellaveglia, M.; Castellano, M.; Di Giovenale, D.; Di Pirro, G.; Ferrario, M.; et al. Strong nonlinear terahertz response induced by Dirac surface states in Bi_2Se_3 topological insulator. *Nat. Commun.* **2016**, *7*, 11421. [CrossRef] [PubMed]
95.	Mitrano, M.; Cantaluppi, A.; Nicoletti, D.; Kaiser, S.; Perucchi, A.; Lupi, S.; Di Pietro, P.; Pontiroli, D.; Riccò, M.; Clark, S.R.; et al. Possible light-induced superconductivity in K_3C_{60} at high temperature. *Nature* **2016**, *530*, 461–464. [CrossRef] [PubMed]
96.	Chen, H.T.; Kersting, R.; Cho, G.C. Terahertz imaging with nanometer resolution. *Appl. Phys. Lett.* **2003**, *83*, 3009–3011. [CrossRef]

MDPI

St. Alban-Anlage 66

4052 Basel

Switzerland

Tel. +41 61 683 77 34

Fax +41 61 302 89 18

www.mdpi.com

Condensed Matter Editorial Office

E-mail: condensedmatter@mdpi.com

www.mdpi.com/journal/condensedmatter